Suburban Planet —

Urban Futures Series
Talja Blokland, *Community as Urban Practice*
Julie-Anne Boudreau, *Global Urban Politics*
Roger Keil, *Suburban Planet*
Loretta Lees, Hyun Bang Shin and Ernesto López-Morales, *Planetary Gentrification*
Ugo Rossi, *Cities in Global Capitalism*

Suburban Planet
Making the World Urban from the Outside In

Roger Keil

polity

Copyright © Roger Keil 2018

The right of Roger Keil to be identified as Author of this Work has been asserted in accordance with the UK Copyright, Designs and Patents Act 1988.

First published in 2018 by Polity Press

Polity Press
65 Bridge Street
Cambridge CB2 1UR, UK

Polity Press
101 Station Landing
Suite 300
Medford, MA 02155
USA

All rights reserved. Except for the quotation of short passages for the purpose of criticism and review, no part of this publication may be reproduced, stored in a retrieval system or transmitted, in any form or by any means, electronic, mechanical, photocopying, recording or otherwise, without the prior permission of the publisher.

ISBN-13: 978-0-7456-8311-9 (hardback)
ISBN-13: 978-0-7456-8312-6 (paperback)

A catalogue record for this book is available from the British Library.

Typeset in 11.5 on 15 pt Adobe Jenson Pro
by Toppan Best-set Premedia Limited
Printed and bound in Great Britain by Clays Ltd, St. Ives PLC.

The publisher has used its best endeavours to ensure that the URLs for external websites referred to in this book are correct and active at the time of going to press. However, the publisher has no responsibility for the websites and can make no guarantee that a site will remain live or that the content is or will remain appropriate.

Every effort has been made to trace all copyright holders, but if any have been inadvertently overlooked the publisher will be pleased to include any necessary credits in any subsequent reprint or edition.

For further information on Polity, visit our website: www.politybooks.com

For the young people (you know who you are).
Condodwellers in a sea of suburbs.

Contents

Acknowledgements — viii

1 Introduction — 3
2 Suburbanization Explained — 23
3 Suburban Theory — 41
4 Suburban Studies — 61
5 From Lakewood to Ferguson — 83
6 Beyond the Picket Fence: Global Suburbia — 109
7 Suburban Infrastructures — 131
8 The Urban Political Ecology of Suburbanization — 151
9 The Political Suburb — 181

Notes — 202
References — 205
Index — 234

Acknowledgements

Work for this book was supported through my York Research Chair and by the Major Collaborative Research Initiative Global Suburbanisms: Governance, Land and Infrastructure in the Twenty-first Century, funded by the Canadian Social Sciences and Humanities Research Council 2010–18 (SSHRC), for which I was the principal investigator. I am grateful to have had the opportunity to lead this initiative over the past seven years and to be able to work with more than fifty researchers and twenty partners on the frontier of global suburbanization. I have learned from them all and I hope they find the seeds they sowed in the pages of this book. I am well aware that I am unable even to begin to represent the rich tapestry of work my colleagues have produced but I did honestly try to distill much of it to the best of my abilities. I am also unable to thank everyone individually here but my gratitude for their having accompanied me on this path is deep. Still, I am particularly indebted to Robin Bloch, who has been my suburban soul brother for more than thirty years, Pierre Hamel with whom I have now collaborated for almost two decades on a variety of projects relevant to this book, and Sara Macdonald with whom I have worked and travelled through the world's peripheries for a decade.

I thank my students at FES and beyond who inspired me. I have had research assistants over the years for some of the work that found its way into this book. Foremost among them was Jenny Lugar who did a marvellous job burrowing through entire fields of suburban study and identifying important writing to me. For the particular manuscript that became this book, Joyce Chan provided formatting and bibliographic help. I also thank the editorial staff at Polity for their professionalism and guidance in bringing this book to fruition.

Sections of this text build selectively on previous or forthcoming published work by the author. These include a paper for Built Environment with the title 'Towers in the park, bungalows in the garden: Peripheral densities, metropolitan scales and the political cultures of post-suburbia'; an article co-authored with Pierre Filion, 'Contested infrastructures: Tension, inequity and innovation in the global suburb' in *Urban Policy and Research*; with Eric Charmes in IJURR on 'Post-suburban morphologies in Canada and France'; and with Sara Macdonald, 'Rethinking urban political ecology from the outside in: Greenbelts and boundaries in the post-suburban city', in *Local Environment*. Some overlap exists with chapters I have produced for John Harrison and Michael Hoyler's *Doing Global Urban Research*, Sage; Henrik Ernstson and Erik Swyngedouw's *Interrupting the Anthropo-ob(S)cene: Political Possibilities in the Natures of Cities*, Routledge; Berger and Kotkin's *Infinite Suburbia*, MIT; and Jayne and Ward's *Urban Theory: New Critical Perspectives*, Routledge.

I have benefited from visiting professorships at the University of Aberystwyth, Université de Montpellier 1, Technische Universität Darmstadt, the Wits City Institute in Johannesburg and the University of Manchester. I have had the opportunity to speak to many audiences across Canada and the world about aspects of this project. I know that the feedback I received there made my thinking clearer. I hope that this translated into the writing, too.

I cannot let this go without acknowledging my Twitter community: Tweeps, you have been an endless font of information, sometimes too much to process, for this long trek I have been on.

Finally, my love goes to Ute Lehrer, at my side in the suburbs of the planet for more than a quarter of a century. If there were any suburbs on the moon, I am sure you'd also come along.

Roger Keil, Toronto April 2017

Suburban Planet

Nova Lima, Belo Horizonte, Brazil

1 Introduction

'[T]he tremendous concentration (of people, activities, wealth, goods, objects, instruments, means and thought) of urban reality and the immense explosion, the projection of numerous, disjunct fragments (peripheries, suburbs, vacation homes, satellite towns) into space.' (Lefebvre 2003: 14)

When it comes to how and where we dwell, work and have fun, we live in times of rapid change. Few periods in history, barring the industrialization of Europe, the urbanization of Latin America and the suburbanization of North America, have seen as much change as the period we are currently going through. We can assume that urbanization marks the moment of our shared experience as planetary citizens. The United Nations' *World Urbanization Prospects* (United Nations 2014) estimates that while in 1950 a total of 746 million lived in urban environments, by 2045 more than 6 billion are expected to be urbanized. This development has been widely understood now to be shaping global development goals in what some have referred to as the Urban Age (Burdett and Sudjic 2007; 2011; Brugmann 2009). Such thinking has been subject to some serious methodological criticism as scholars have pointed out that we ought to think less of people in cities than people in urban society, less in categories such as global and megacities and more in terms like 'planetary urbanization' (Brenner and Schmid 2015; Gleeson 2014; Ren and Keil 2017).

While these critiques are incisive and important, the current book aims at an intervention on a different terrain. The notion of an urban age suggests in its core a move of urban populations from more dispersed into denser environments for residence, work and recreation. This move towards more compact spatial patterns for work and life is certainly borne out by the world's 'final migration' to move to the 'arrival cities' of the twenty-first century (Saunders 2011). The global migration of millions of first time urbanites in less developed countries is mirrored by a distinct move towards re-urbanization in industrialized countries that had been going through half a century of de-industrialization, suburbanization and urban decline. What is more, these processes have been welcomed and normatively prescribed by planners and urbanists responding to challenges of climate change and sustainability that are said to be met more readily in compact, denser cities. While such processes of re-urbanization, densification and compactness are real and imagined features of the urban age, this book occupies itself with questions of urban growth that are better understood if we take into account tendencies towards urban expansion, de-centralization and suburbanization. As we will explore in a sequence of historical, conceptual and thematic chapters, much of the urban age is, at closer inspection, rather a *sub*urban age. We live on a suburban planet. This observation is supported by statistical evidence that shows, as has the work of Shlomo Angel and colleagues (Angel, Parent and Civco 2010; Angel, Parent, Civco and Blei 2010) among others, that the growth of cities' populations and activities is characterized by a disproportional expansion of those cities' territory. In other words, as the world urbanizes, cities also become less densely populated, their spaces less intensively used.

It is expected that in 2030 urbanized land on the planet will cover 1.2 million square kilometres. That is twice as much as in 2000. Urbanization at this incredible rate must give everyone pause. This ubiquitous trend will imply significant consequences for climate change, biodiversity and so forth (Seto, Güneralp, Hutyra 2012). In the near future, another three billion humans will have to be housed.[1] Most of the future inhabitants of the earth's crust will be living in entirely new cities that run the spectrum from Corbusian nightmares to 'broadacre' campuses and squatter camps and many more of those who already live in cities will move to those new, mostly suburban environments, too. None of them, perhaps with the exception of the odd historicizing Chinese new town, will look like nineteenth-century Manchester or Paris, or the heart of Amsterdam or Barcelona (Swilling 2016). Two aspects stand out in this perspective: first, this projected urbanization will be extremely unequal, with China and Africa absorbing the lion's share of global urbanization during the next generation; second, we can expect that the majority of the urban expansion we face in the next generation or two will not mirror the current trend towards re-urbanization, widely celebrated in the urban North, but will continue to be extensive in nature. This will take wildly different forms in places such as China or Turkey, where more dense, high-rise type suburban developments are driven by large-scale state-sponsored programmes and (most of) Africa or India, where we see continued and continuous lower density suburbanization prevail (Bloch 2015; Gururani 2013; Mabin 2013; Wu and Shen 2015; Wu 2013). At the same time, there will also be additional suburban extension of cities in North America, Europe and Australia, where half-hearted growth controls can barely withstand the tide of further sprawl – residential, commercial and industrial –, often now driven by aggressive infrastructure development, including airports, private motorized transportation and public transit that now reach the far corners of the commuter shed and extend the urban region (Addie and Keil 2015).

The notion of an urban planet is not new. Manuel Castells, for example, notes as early as 1976 that American elites were operating on the assumption that an urban world had dawned. He quotes Senator Abraham Alexander Ribicoff, who observed : 'To say that the city is the central problem of American life is simply to know that increasingly the cities are American life; just as urban living is becoming the condition of man across the world....' (quoted in Castells 1976: 2–3). Ribicoff emphasizes a qualitative, rather than a mere quantitative shift towards urban life: 'The city is not just housing and stores. It is not just education and employment, parks and theaters, banks and shops. It is a place where men should be able to live in dignity and security and harmony, where the great achievements of modern civilization and the ageless pleasures afforded by natural beauty should be available to all' (quoted in Castells 1976: 3). Henri Lefebvre, had preceded Castells by a few years to announce the coming of an 'urban society'. In his book *La revolution urbaine*, first published in English in 2003, Lefebvre predicted that society was going to be irreversibly urban as the production of urban space was becoming the central process through which capital was accumulated and the reach of the city extended far beyond its immediate physical borders through metabolic relationships that spanned the world (Lefebvre 2003).

In the half-century that has passed since these early premonitions of an urban world, many have jumped on the bandwagon of pronouncing the onset of the urban age. An entire industry of Lefebvre scholarship has sprung up to celebrate the possibilities of claiming the Right to the City in an urbanized world. Most recently, perhaps, the declaration of an epoch of 'planetary urbanization' has been the most influential mode of carrying Lefebvre's message forward (Brenner 2014). In this wave of urban enthusiasm for the revolutionary potential of urbanization and sharp critique of mainstream urbanism, the idea that cities both 'explode' and 'implode' as expressed in the epigraph at the top of this chapter has gained ground to describe that processes of densification 'here' stand in direct relationship with processes of de-centralization 'there'. An 'oscillating growth dynamics' (Keil and Ronneberger 1994) has been characteristic of the waves of urbanization that have swept the world's metro areas: as the centres gain in population and economic activity, so do the peripheries. When the Frankfurt Airport at the city's edge expands, so do the activities of the financial industry in the core. As the fringe of Toronto is stretched into the fields abutting the protected Oak Ridges Moraine (Gee 2017), the condominium towers in the inner city mushroom into the sky. As Los Angeles makes one attempt after another to anchor a resident population in its gentrifying downtown (Dillon 2017), its desert frontier continues to be pushed out indefinitely despite the (subprime) crisis writings on the wall.

John Friedmann, the great American planner, has used the terms 'prospect of cities' and 'the urban transition' to point out the irreversible inevitability of the world turning urban (2002). The urban transition is the unstoppable movement from the rural and agricultural to the urban. Already in the 1960s, he saw the urban field as the spatial form of the urban transition: 'It is no longer possible to regard the city as purely an artifact, or a political entity, or a configuration of population densities. All of these are outmoded constructs that recall a time when one could trace a sharp dividing line between town and countryside, rural and urban man.' (Friedmann and Miller 1965: 314). This transition is, of course, not just one of brick and mortar, infrastructures and technologies, it involves urbanization of the world's ways of life: 'We are headed irrevocably into a century in which the world's population will become, in some fundamental sense, completely urbanized' (2002: 2). The expansion and consolidation of global capital can be assumed to continue and the urban transition to be completed over the course of this century. Friedmann and Miller note in their once pathbreaking piece on the urban field: 'The pattern of the urban field will elude easy perception by the eye and it will be difficult to rationalize in terms of Euclidean geometry. It will be a large complex pattern which, unlike the traditional city, will no longer be directly accessible to the senses' (1965: 319). This expansion of urban form and urban imaginary is akin to the explosions noted by Lefebvre (2003).

In *Time Magazine* Bryan Walsh (2012) notes that 'The urbanization wave can't be stopped – and it shouldn't be' but pleads for a change in the way we build cities in the future in order to make urbanization sustainable. But how are we really building cities? It seems that over the past two hundred years, not to mention for thousands of years before, we have been mostly building cities in a manner of concentric expansion of rings around a typically religious or secular seat of power and around a market place. Many core cities have experienced 'Haussmannization' of one kind or another throughout the twentieth century, with added axes, corridors and central intensification and structuration. Others have been altogether invented as products of a modern age, as have Brasilia or the bombed out European cities after the Second World War. But most of modern urbanization has been an ongoing process of suburban extension. Over time, those suburbs became cities themselves. They tended to become denser, less informal, more reliant on technologies of mobility. In pre-industrial times, the extension was minimal. Much of it had to do with limited technologies of mobility. Militarily the city's compact form was also an advantage as it could be defended better. In the industrial and automobile city of the twentieth century, the extension is celebrated not just as a temporary state that would soon be giving way to a more traditional form of urbanism. On the contrary, the extension was now considered a legitimate, if not preferred form of urbanism: in the periphery, more than in the centre, the true expression of society's success was to be found, whether that was the single-family home subdivision of the American dream, the new towns of the Soviet empire or the housing estates of social democratic Europe and Canada. That there would be a way back to the core from the garden cities, satellite towns and subdivisions of the automobile metropolis would have sounded implausible or even undesirable to mid-century planners and city builders. Yet, for the most part, those consecutive waves of twentieth century suburbanization were still dependent on and related to the urban centre, whether it was for financing of infrastructure, jobs or government services. Towards the end of the millennium, though, this changed. Often discussed in the context of the Los Angeles School of urban studies, we are beginning to see the dissolution of centrality as we knew it. Instead, the urban form becomes polycentric and the suburbs themselves appear more as free-floating units (Soja 1996; 2000).

The argument I am putting forward in this book is that under the conditions of current trends in technology, capital accumulation, land development and urban governance, the expected global urbanization will necessarily be largely *sub*urbanization. Yet, we (as in urban professionals, planners, scholars of urban studies, etc.) have collectively embraced centralism and compactness as the guiding idea of twenty-first-century urbanism, just at a time when there is a massive explosion in the way land around existing cities is used. Suburban land, as one of the chief products of post-neoliberal capitalism, continues to be readied for settlement whether it is in the form of subdivisions in North America or squatter settlements in India or Africa. Around the globe, *sub*urbanization now occurs without the automatic assumption that this may lead to denser, more central forms of urbanization later – although it may.

Following Andy Merrifield's (2012) call for a new conceptual register beyond the traditional dichotomies of urban studies and for a re-theorization of urbanization more fundamentally, this book on global suburbanization is by no means intended to reify and mark differences between the category of 'suburb' and the rest of the dimensions through which general urbanization moves ahead. The traditional dichotomy of city and suburb has itself presented an obstacle to a better understanding of urbanization overall (Schafran 2013). Yet it has also been at the core of what we take to be the difference between the good city and the bad, which Alex Schafran has called an 'ongoing duel between utopian and dystopian visions [that] has been constructed discursively, at least in the American context, through the defining dialectic of the past century of American metropolitan thought – the city versus the suburb' (Schafran 2013: 136). In this sense, the book understands its place to be part of what Merrifield (2012) names a 'reloaded urban studies' that dispenses 'with all the old chestnuts between North and South, between developed and underdeveloped worlds, between urban and rural, between urban and regional, *between city and suburb* and so forth'. Therefore, the book is as much a specific intervention into suburban debates as it is a contribution to a rejuvenated conversation on urban theory overall (see, for example, Judd and Simpson 2011; Robinson and Roy 2016).

This, then, is a book about global suburbanization as a particular and pervasive process of urban expansion. That said, let me do away with a few misconceptions right at the start. This book is not a normative plea for suburbanization. While ostensibly about suburbs and suburbanization, it is not a book *against* cities and urbanization. Recognizing the sustained and perhaps growing significance of suburban forms of life and peripheral modes of urbanization, does not mean discounting the remarkable push of development towards the city centres, the tendencies towards re-urbanization and the recovery of urban cores in many parts of the world. Much of that inner-city regeneration is tied into waves of gentrification and expulsions of poorer populations from city centres as a new spatial fix – that of a creative knowledge economy – is taking hold in the de-cored manufacturing metropoles of the past.

So, this book is primarily about Lefebvre's 'explosions' that are part of the urbanization processes we are currently undergoing. The metaphor of 'implosions/explosions' is itself marred by severe limitations due to its particular time-specific use in the age of astrophysics but for now, let's assume that suburbanization is part of that centrifugal movement that creates unstructured communities of usually lower densities beyond the classical core of the city that both contracts and increases in height and scale. Yet, when we emphasize the importance of suburbanization and while we are doing this without denying the continued or even increasing metropolitanization and re-urbanization that occurs at the same time, we are also not talking about the kind of suburbanization that has traditionally given the process its name. In order to follow me down the road of my argument in this book, I will at least temporarily and partly have to ask the reader to suspend images of white picket fences and single-family homes on cul-de-sacs when reading the words suburb, suburbanization and suburbanisms.

What do I mean by sub/urbanization and sub/urbanism more generally and suburbanization and suburbanism more specifically in this book? In the first instance, suburbanization 'is a combination of non-central population and economic growth with urban spatial expansion' (Ekers et al. 2012: 407). Suburbanism(s) refers to suburban ways of life. The definitional simplicity must not overshadow, however, the vast diversity of processes and forms we find in suburbanization and suburbanisms worldwide. Importantly, then, this book is not about suburbs as things but about suburbanization as a process. Suburbanization is a process of active production and reproduction. This includes discursive processes of world production. A process of worlding in the sense of being the process that makes the world today. The trope harkens back to what David Harvey said about the need to focus on urbanization as a process instead of the city as a thing (1996). This book attempts to help recognize the role of the periphery in these processes. Specifically, suburbanization is seen as a product of self-built, state-led and private-led development; these three styles can and mostly do exist simultaneously and in combination and are not to be understood as occurring in historical sequence (Ekers et al. 2012). This leads to an insight that is fundamental to the methodological approach that underlies this book: 'In contrast to periodizing suburban expansion and decline, distinguishing between self-led, state-led and market- or private-led development, avoids taking the Euro-American case as fundamental and highlights divergent yet comparable processes in different spaces' (Ekers et al. 2012: 411).

While other books, approaches, views are seeing 'a world of suburbs' (Harris 2010), this book sees the world through suburbanization. Suburbanization is not an epiphenomenon of the way the world evolves/revolves. It is the very looking glass through which we see the world today critically. As Lefebvre's urban revolution turns, the suburbanization of the city region takes over the planet. Whatever view one holds of urbanization processes – and I am taking a relational, topological view here – the reality of life as we know it today is marked by multiscalar everydayness (*Alltäglichkeit*). This book is not a short history of suburbanization and not a systematic review of suburban studies. It is also not an atlas of suburbanism. A large number of recent publications provide those aspects of the study of suburbs (Moos and Walter-Joseph 2017). What the book provides instead is a set of arguments about suburbanization and suburban ways of life. The very concept of suburb or suburban has recently received renewed attention. Some guidance will be obtained from work by authors who have attempted to distil 'meaningful types in a world of suburbs' (Harris 2010). Taxonomies and lexicons of suburbanization have been developed. 'The suburb' has been in the centre of these considerations (Harris and Vorms 2017). Building on but also in contrast to these important contributions, this book attempts a less defining and more inquisitive approach. While less interested in laying out the conceptual boundaries of a thing called 'suburb' I am keen, instead, on contextualizing the continuous suburbanization of our world in a general project of urban theory building.

As for the specific proliferation of the global suburban, there is some convergence: tracts of single-family suburban homes behind gated walls; malls and freeways and airport warehousing landscapes; flood control infrastructures that concrete regional waterways into suburb-compatible flows and streams; edge cities with offices and condominiums, etc. But the real domain of convergence in form and function remains the inner city with its shared ambitions of creativity, capital accumulation and culture and its hyper-gentrification that flattens all difference across the globe. By contrast, the range of suburban and post-suburban developments we can register is filled with surprises throughout. Whether we look, for example, at the rebuilding of suburban modernist tower neighbourhoods in the East and West, the bustling horizontal slums and squatter settlements as well as new middle-class neighbourhoods of African urban peripheries (Bloch 2015; Mabin 2013), the vertical suburbs of Chinese megacities (Fleischer 2010; Wu and Shen 2015), the 'classical' Levittown suburbs, the foreclosure-ridden 'vulgar' exurbs in the deserts and woods of the USA (Knox 2008), the transition towns in English greenbelts, edge cities, ethnoburbs or any other form of suburban settlement, there is more diversity to be found there than perhaps anywhere else in the modern history of city-building and re-building.

This helplessly incomplete list also reveals that this book is not about a particular type of suburb, in the least about the North American suburb which has long been considered the model case for the phenomenon and has caught the majority of interest among writers and thinkers on the subject. While the single-family home residential suburb of the post Second World War era in the United States (and to a degree Australia, Britain, Canada) gets its share of attention in this book, its specific problematiques will not be a stand-in for the suburbanization of the world or, to invert this notion: globalized suburbanization. The book presents a state-of-the art sketch of critical thinking on the suburban question and pushes beyond the *empirical* and *conceptual* evidence presented by the world's leading thinkers on suburbs into the speculative terrain of new theorizing on the urban more generally. In doing so, the book takes a stance much different from most of the existing literature that sees suburbs as derivative (of the 'city'); as problematic and lacking (as in social life and environmental sustainability) and as uniform (as in built form). Much more, the book gives the suburbs their rightful place in the conceptual imaginaries and real worlds through which we must understand the global urbanism in which our lives now are being lived.

The book is naturally critical of the conventional suburb-boosting arguments emanating usually from American libertarian scholars and pundits. They deserve their day in court but there is not much here to pursue in good faith. Yet it is also not another tired tirade against the suburbs as the place of all things wrong with today's cities (cars, malls, boredom, sprawl, etc.). While not mincing words on the toll exerted by suburbanization on human and natural environments, the book explores the suburbs without moral judgement applied. Calling urbanization peripheral is just a starting point for a larger discussion about the changing urban geographies of centre and periphery. In debates as different as classical rent theory, Chicago School urbanism or critical urban theory (most notably of the kind influenced by Henri Lefebvre), centrality is often taken literally. Suburbanization, then, has historically been seen as a move away from the core of society and city in more than just a spatial sense. This book will interrogate this shift from centre to margin and discuss 'the right to the city' in light of recent dynamics in urbanization.

Much or even most of what we see today as the building, re-building and un-building of cities is suburban. Moreover, much of urban expansion and urban change (including shrinkage) occurs in a *post-suburban* environment where classical suburban expansion is just one in a range of ways in which cities' peripheries are being reformed and rearranged. Complex post-suburbanity, more than just primary or original suburbanization, is the topic of this book (Phelps and Wu 2011).

The book is not about any specific suburb. While experiences and data from many real places inspire and fuel the narratives in this book and while I will strongly rely on my own research and lived experience over the past twenty-five years in Frankfurt, Los Angeles, Toronto and Montpellier, no particular place/case studies are being presented here.

The book will return periodically to three areas of interest: physical form/built environment; social relationships/process/governance; (sub) urban political ecologies. Materially, the book orients itself through a critical look at the governance, land and infrastructure of suburbanization as three relevant dimensions through which real suburbanization proceeds. Following Ekers, Hamel and Keil (2012) I pursue the production of suburban space as a combination of state, capital and private (often authoritarian) action.

The book is not meant to be an encyclopedia of suburbanization. Rather, it is a long essay on the subject. Respectful of the work done so far in urban studies on the subject of suburbanization, *Suburban Planet* pushes beyond the 'state of the art'. While subjective and argumentative, it is based on information from perhaps the largest globally scaled research project on suburbanization and suburbanisms: 'Global Suburbanisms: Governance, Land and Infrastructure in the twenty-first Century' of which I have been the principal investigator. This Major Collective Research Initiative housed at York University is funded by the Social Sciences and Humanities Research Council (SSHRC) and lasted from 2010 to 2018. Empirical research from this project and other foundational work by likeminded researchers will be processed and critically engaged with through the chapters of this book as appropriate.

In short, the book will argue that:

- The urban century is really the suburban century. In a majority urban world, most activity in terms of the expansion and contraction of urban population, built form and economic activity will occur in peripheral areas.
- As the world continues its 'urban revolution' (Lefebvre 2003), attention needs to be paid to how this is related to the specific processes of spatial, economic and social peripheralization characterizing urbanization today.
- Whether we call these forms suburban, peri-urban, peripheral, post-suburban, exurban or employ any equivalent concepts, the phenomenon, to borrow from Lefebvre, 'is universal' (Lefebvre 2003: 54).

- It is through the lens of suburbanization (broadly defined) that the process of urbanization reveals itself in the twenty-first century.
- The book will provide a strong argument about the possibility of an expanded comparative perspective on global suburbanization and to move away from the 'classical' case of the North American suburb as the assumed benchmark for suburban studies.

WHAT KIND OF WORLD DO WE LIVE IN? A WORLD OF SUBURBS!

The global suburban landscape now has a kaleidoscopic appearance. There is great multiplicity in the rapidly suburbanizing geographic regions all over the world. But this multiplicity is also somewhat deceptive. There is much blurring and bleeding among and between the different world regions. In a post-colonial, post-suburban world, the forms, functions, relations, etc. of one suburban tradition get easily merged, refracted and fully displaced in and by others elsewhere, near or far. Our optics have changed accordingly and we have collectively been challenged to abandon historically privileged spots for observing urbanization. That includes both the privilege of the urban centre and the privilege of the Global North, long considered – and inherently treated – as the norm in trajectories of global urbanization.

I am profoundly respectful of the ultimate unknowability of the urban and implicitly of the suburban. In David Mitchell's 2004 novel *Cloud Atlas* Luisa, one of the protagonists, moves through a fictional city – modelled perhaps on San Diego or Los Angeles in the 1970s:

> Angry horns blast as Luisa fumbles with the unfamiliar transmission. After Thirteenth Street the city loses its moneyed Pacific character. Carob trees, watered by the city, give way to buckled streetlights. Joggers

do not pant down these side streets. The neighborhood could be from any manufacturing zone in any industrial belt. Bums doze on benches, weeds crack the sidewalk, skins get darker block by block, flyers cover barricaded doors, graffiti spreads across every surface below the height of a teenager holding a spray can. The garbage collectors are on strike, again, and mounds of rubbish putrefy in the sun. Pawnshops, nameless laundromats, and grocers scratch a lean living from threadbare pockets. After more blocks and streetlights, the shops give way to anonymous manufacturing firms and housing projects. Luisa has never even driven through this district and feels unsettled by the unknowability of cities. (Mitchell 2004: 419)

The impossibility of really knowing cities that disturbs Luisa stems primarily from our habit of following well-known paths when we move through the urban field. We don't know the neighbourhoods where we don't usually live, work and party. That we are not familiar with those is the result of complex imbricated influences of the deliberate sequestration and self-segregation, of the exclusion of certain spaces for reasons of social and economic difference, as an outcome of internalized ways of life.

In the area of our spatial perception, we usually remain on our normal terrain, in our comfort zone, where we know our way around. Fictional Luisa has left this zone, as she finds herself driving through the suburbs of fictional San Diego. But also the conceived space, identical often with the corporate and state and planning space of capitalist accumulation, confronts Luisa as alienated, because she fails to understand, the fetishized causal processes that lead to social segregation, de-industrialization, to state retreat and so forth (Lefebvre 1991). We can add that Luisa's gaze betrays a focus on the normal and ordered life in the coherent city, which she misses in the impoverished and de-industrialized suburb. But Luisa's individual experience of ignorance (or alienation) plays into the general insight that these complex processes of urban creative destruction in the post-suburban age cannot be understood as long as we train the categories of the familiar onto the unfamiliar, i. e. approach the rapidly changing instances of sub/urbanization in the twenty-first century with the perspective of the nineteenth and twentieth centuries. During the nineteenth century, urbanization and industrialization were considered as one in the industrializing European and American nations where city dwellers began to make up about fifteen per cent of the population by 1900; in the twentieth century, metropolitanization went along with the massive industrial shift that included rapid technological change, mechanization, automobilization, mass production etc. from which emerged the regionalization of the city and its apparent 'dissolution'. At the beginning of the twenty-first century, a bifurcation ensues which sees on one hand the re-urbanization of the sprawling metropolis – which Alan Ehrenhalt (2012) has called 'the great inversion' in North America. To some degree, we can also observe this 'inversion' in Europe – which has traditionally kept its wealthy downtown and has exported its poor to the periphery – where now the hyper-gentrification of the city cores is accompanied by a massive decomposition of urban order in sprawling inbetween cities, or *Zwischenstädte* (Sieverts 2003). On the other hand, we see new modes of metropolitanization, suburbanization and de-urbanization in Africa and to some extent in Asia that don't follow the assumed trajectory of urbanization known to Europeans and Americans since the nineteenth century (Phelps and Wu 2011; Mabin, Butcher, Bloch 2013).

The literary example from *Cloud Atlas* also points to further connected problematiques of sub/urban ontology and epistemology: at the start, this concerns our capacity to have knowledge about the city. The predominant question here is in how far we, as users and everyday producers of urban ways of life, can know the city beyond our own experience. Of course, experience is a limited and limiting category of knowledge acquisition. Limited because it is subjective and never encompassing. In the Lefebvrian sense, the experience is 'space of perception' and the dimensions of the conception and of the lived space are not accessible through it. Although impressions gained from art, literature, music, documentation, etc. that the urban dweller carries with him- or herself belong to the rich experience that influences perception, the process remains subjective. Scientific analysis, on the other hand, builds on hedging in the subjective, experimental, personal dimensions of knowledge, in favour of codified and normed forms of the appropriation of reality.

Experience is always limiting since one's own view might block access to knowledge on other, unknown aspects of the urban. The bourgeois prejudices on urbanity align with certain spatial images and are taken for granted after a while. Beyond these assumed categories of the urban, openings are rarely created for us to know the city in ways that are different from what the canon would suggest. In Europe, this means that the urban is taken as a given for the inner cities, while it is denied to the suburbs. In the USA, the urban exists as a dreamworld and world of nightmares beyond the suburbs, which act as the privileged position from which 'the city' is experienced. But here also, the racialized and classed perspective inherent in this position blocks a more comprehensive view of the city as a form of life. In this sense, then, suburbanization and suburbanism appear as an epistemological filter. Their *felt* specificity and their *assumed* less-than-urban status present obstacles to suburbanites and non-suburbanites alike, to seeing the urban for what it is: a place of contradiction and spatialized plurality (structured and segregated by state power, market dynamics and private authoritarianism) in which both centrality and peripherality are related values on a continuum and antitheses in a dialectics of ongoing urbanization.

My argument in this book follows a suggestion by Ananya Roy to approach urbanism in a four-dimensional manner. First, 'urbanism refers to the territorial circuits of late capitalism' (2011: 8). Privileging the political economy of land production in our analysis is a critical statement in an environment where often lifestyle choice and consumer privilege is seen as the driving force of urbanization. Second, there is the acknowledgement that while capital tends towards structuring urban space in its image, it is unable to structure it at will. In contrast, 'urbanism indicates a set of social struggles over urban space'. Here we deal with the claim to the right to the city, or as we will see, the right to the suburb. Third, we can discuss urbanism as a 'formally constituted object, one produced through the public apparatus that we may designate as planning'. While this book is not about planning per se, this includes the recognition of state action, coordinated effort and contested processes of rule-making around the production of urban space. And fourth and lastly, Roy notes that 'urbanism is inevitably global' (2011: 9). This is the pervasive theme of the narrative presented here.

Looking at suburbanisms as global phenomena creates the uneasy necessity of having to step out of the secure space of national and known trajectories of urbanization in regions and cities and out of the safety zone of positivist linearities of urban growth (and decline). Rather, following Roy, we are adding 'unthinkable space' to the already established term 'unknowability' of the city. For that, Roy borrows from Derek Gregory and describes this space as 'seemingly unplanned, seemingly undecipherable, marked by unimaginable fragmentation and extraordinary violence…that may speak to some of the most prevalent urban and thus human conditions of the twenty-first century' (2011: 9).

Roy uses yet another operative term in her analysis that becomes a key methodological concept for understanding the urban process today. It is the notion of 'worlding'. She develops this notion from the insight that not all cities are following a preordained path with declared outcome: 'In contrast, the concept of worlding seeks to recover and restore the vast array of global strategies that are being staged at the urban scale around the world' (2011: 9).

The chapters that follow take up the topic of global suburbanization through a variety of conceptual and empirical lenses. Chapters 2–4 look at how suburbs have been explained, theorized and studied. Chapters 5 and 6 examine the changing composition of the traditional suburb and the proliferation of the suburban around the world. Chapter 7 discusses suburban infrastructures. Chapter 8 explores suburban political ecologies through the lenses of density, boundaries and the Anthropocene. A chapter on the political suburb concludes the volume.

Stockholm, Sweden

2 Suburbanization Explained

Suburbs are often thought of as residential enclaves, conventionally middle class and consumerist and in contradistinction to the complexity of the central city with its multiple diversities and contradictions. This common image of suburbia is reflected in much of the literature on peripheral urban development. In one of the standard works in suburban studies Harris and Larkham (1999: 8) define suburbs as peripheral in location vis-à-vis a centre, residential, low density, a distinctive way of life and municipal political autonomy. As Nick Phelps has pointed out, such a definition is not plausible in light of 'the outward expansion of urban areas [that] renders distinctions between city and suburb rather arbitrary' (2012: 259). The focus, in particular, on suburbia's residential nature and its low-density form must be called into question. Robert Lewis points to the bias in suburban studies that has contributed greatly to a particular view of what suburbanization is all about: the privileging of middle-class residential environments, close to nature, or in nature, usually on tree-lined street systems with an internal orientation for private consumption and external links to higher-level transportation (transit and autoroutes) that connects to labour markets, cultural and educational facilities downtown as well as commercial establishments (malls, entertainment centres, etc.) in nodal subcentres along the extended mobility grid (Lewis 2004b: 3). The history of North American suburbanization has largely been written with a decided emphasis on the role of residential spaces. The literature presents much by way of post-hoc description of the process and outcomes of a century of suburbanization in North America but comparatively less by way of analysis and explanation. What, then, do we need to know about how and why it all came about?

Lewis himself offers some insight into where we might look: one of the consistent ways in which suburbanization – both residential and industrial – has been explained is in a Marxian inflected literature on metropolitan growth (2004b). Richard Walker (1981), for example, most explicitly linked suburbanization to accumulation cycles in the capitalist city and pointed to the peripheralization of investment as a 'suburban solution' to crisis and overproduction cycles, a form of capital switching. Following this theory, suburbanization was a particular 'spatial fix', that offered both short-term relief for capital investment into the built environment of the urban periphery and long-term virtual cycles of investment into a suburban landscape that longed for the kinds of commodities and gadgets that came rolling off the Fordist assembly lines in the post Second World War period. The often overlooked industrial suburbs had yet an added, non-economical, but rather political rationale built in: 'Changes to urban form and industrial and residential suburbanization were structured by class conflict and emerged as a solution to capitalism's internal contradictions. The corporate city, for example, was defined by the de-centralization of industrial activity as capitalists sought factory sites away from worker unrest in the central city' (Lewis 2004b: 5). This analysis of suburbanization as part of the conflictory and contradictory metropolitanization of capital is crucial to most of the critical literature on suburbanization to date. It avoids two important flaws of the standard literature on suburbs: (1) It takes the suburbs as a product of capital rather than individual market choice; and it (2) looks at suburbs not as a sequestered and idealized space beyond the city's borders but as an integral part of larger processes of urbanization and metropolitanization throughout the past long century.

It cannot be emphasized too much that the focus in suburban studies on residential choice rather than industrial planning has skewed the way academics and public opinion have viewed the suburbs over time. Industrial suburbs have not been a late addition to what essentially was a residential periphery (Walker and Lewis 2001). Quite in contrast, industrialization and suburbanization have been linked since the early days of urban expansion in the nineteenth century. In a collusion of geographic economic clustering, the political economy of (sub)urban land and public policy, industrial suburbanization has been one of the chief trends of metropolitanization for one and a half centuries (Harris 1996). For Walker and Lewis, this insight does not just mark a different set of historical facts than the ones traditionally taken for granted, they also provide the basis for a reevaluation of suburban and maybe urban theory altogether (Walker and Lewis 2001: 7).

Suburbanization of housing and industry that has traditionally been seen as secondary to the urbanization process gains in significance and characterizes the urban and the social more than ever before. With this, we encounter a central and recurring theme of this book: the long, but recently accelerated dialectics of sub/urbanization, a story often told and retold as a narrative of loss (of the centre, the city itself, the future, etc.). Henri Lefebvre famously said at the end of his life: 'There was a time when city centres were active and productive and thus belonged to the workers (*populaire*).'[1] In this epoch, moreover, the City (*cité*) operated primarily through the centre. The dislocation of this urban form began in the late nineteenth century resulting in the deportation of all that the population considered active and productive into suburbs (*banlieues*), which were being located ever further away. The ruling class can be blamed for this, but it was simply making skillful use of an urban trend and a requirement of the relations of production. Could factories and polluting industries be maintained in the urban cores?' (Lefebvre 2014 [1989]: 567).

In more recent times, of course, economies of the centre have been imagined in different terms. Instead of the 'polluting industries' of yore, the so-called creative sectors of the economy, the cultural industries in particular, have received most of the attention of economic geographers and the general public. The prioritization of the cultural industries and the creative came largely with a rediscovery of the centrally urban as a productive locale: dense inner cities have since been seen as natural habitat and workplace of a generation of 'millennials' who turned their back on their parents' and grandparents' suburban occupations and residential choices. The urban bias of Jane Jacobs's (1969) recently resuscitated economic writings and the popularization of Richard Florida's (2002) work about 'the creative class' has contributed to the preferential view of downtown spaces for the talented professions who are considered the drivers of a re-tooled capitalist urban economy. It is noteworthy here, as Phelps (2012) has pointed out, that in this scheme the suburbs are viewed primarily as a 'sub-creative' place where innovation and productivity (and creatives) are not at home (see also Lawton et al. 2013). This perspective has engendered a widely shared perspective that suburbs have never been and are not currently the site of innovation and creativity. Phelps's critique places 'the suburban economy in all its variety (innovative or indeed not so creative) within a larger context of complex and continually evolving, extended metropolitan economies' (2012: 268). This releases suburbia from the margins of the spatio-economic imaginary. [2]

The new importance of the periphery must be noted by critical urban research because many of the points of friction and conflict that characterize today's social and urban life will appear at the *frontier* of the urbanization process in future. At the margins of the metropolis, more than in the increasingly uniform and normed inner cities, new urban forms and ways of life emerge. Those tend to be indicative and directive for our existence in the twenty-first century overall. The argument for paying more attention to the suburban as a way of life and to suburbanization as a process is not an endorsement of the suburbs as places, the mechanisms that produce and sustain them, for the people that populate them and their ways of life and the particular future they signify. It only entails the insight that it is at the spatial margin of urbanization, in the increasingly non-uniform suburbs of the large metropoles, middle cities and even rural towns, where many dramas of the 'urban century' will be played out. The 'urban revolution' which Henri Lefebvre saw coming in the 1960s realizes itself today in the periphery of the urban. Lefebvre notes: 'The city affirms its presence and bursts apart' (2003: 108). Next to centralization, therefore, suburbanization, is one of the foundational pillars of urban society and of what Lefebvre calls the 'planetary nature of the urban phenomenon' (2003: 113).

Today's realization of urban society as predicted by Lefebvre cannot be thought of as the classical formulation of the suburban deficit anymore. Suburbs are not catching up. They determine the pace and direction of urbanization today. The urban revolution does not just return from the periphery to the core but opens the city towards urban society. This means that the full realization of the urban promise, the complete saturation of the social with forms of urban life, now occurs mainly through the explosion of the city into heretofore unknown dimensions.

In some ways, the history of urbanization has always been suburbanization. The expansion outward of historical city cores from the religious-state-commercial-residential centre, the extension of cities in the wider sense, has always been characterized by spatial and functional diffusion. The 'old city' of antiquity and of the European Middle Ages was enlarged, in the modern period, beyond the enforcement walls, after they were dismantled. In this sense, suburbanization appears as the foundational principle of urbanization overall. Urbanization is hard to imagine through thousands of years were it not through the settlement of new land at the urban edge. 'New' land, of course, it really is not. It has always been there. And it is also not empty, not a void. In many cases, it has been under agricultural cultivation for as long as the city that it surrounds. In modern settler societies, like in the Americas or in Australia (Johnson 2013, 2014), land has had pre-colonial histories of indigenous use that may not be urban or even agricultural in nature but may be more ephemeral which makes it hardly 'new' or 'a void' waiting to be urbanized (C. Harris 2004; Jacobs 1996; Veracini 2011).

But the land is 'new' in the sense that it has to be made, has to be produced which, in particular in the modern capitalist period, is a process structured by formal and informal land markets and directed (more or less forcefully) by the state. In the age of financialization, the land game is 'played' by a variety of actors from individual families to global corporations. Suburban land, 'at once transitional and transitory' (R. Harris 2013), is simultaneously product of and precondition for activities of actors in the land market. It has to be first produced – conventionally by turning rural into urban land – and then built upon – a process of wide-ranging variety and multiplicity in form, function and use (Lehrer, Harris, Bloch 2015; Haila 2015). The expansion of land markets has been largely synonymous with the growth of the city and its suburbs, whether that we look at the industrial city in nineteenth-century Europe, the American metropolis of the twentieth century or African or Asian periurban expansions of the twenty-first.

And this growth is not 'value free'. It is always connected to specific groups of humans and processes that drive suburbanization (Logan and Molotch 1987). In the past, often processes of displacement and exclusion led to suburbanization. Jon Teaford reminds us that it was predominantly the unwelcomed functions, people and activities that were once pushed to the periphery and beyond: 'Sprawling livestock yards and brickfields as well as noxious slaughterhouses have been forced to operate on the edge of cities. In the pre-modern era they were joined by the immoral or illegal pursuits of prostitutes of unlicensed hucksters expelled from the walled core' (Teaford 2011: 15). All those unwanted uses that get sedimented at the periphery remain characteristic of the post-suburban landscape of today, be it in the 'sexscapes' of modern suburbia (Maginn and Steinmetz 2015) or the nooks and crannies of the in-between cities where explosive, unhealthy, discarded, unsightly uses abound (Keil and Young 2009).

At the other end of the social spectrum wealthy patricians and the bourgeoisie spent at least some of their time in the countryside in order to escape the effects of dense urban life that were considered negative. Disease and social unrest were suspected to have their natural home in the city, while the countryside was idealized as a pure idyll. The mere temporary move to the suburb has not kept the urban elites from maintaining control over the central symbolic and material spaces of the urban – and to expand that grip through constant processes of gentrification and symbolic occupation with mega-projects, waterfront renewal, spectacle and so forth.

In the urban history of the waning nineteenth and early twentieth centuries, suburbanization gains a higher significance. It develops from necessary expansion to planned invention of a novel urban form that was articulated not with classical notions of privilege – as has been the case with the mansions of the landed gentry – or exclusion – as in the *bidonvilles* of the early modern city. Suburbanization then was part of the reproduction of capital at a scale previously unknown. The production of space and the accumulation of capital became one and the same during much of the twentieth century. A particular form of development in the United States set the pace and created the imaginary for this push. In the Anglo-American countries – Britain – as well as in the capitalist settler societies – Australia, Canada, the United States and to a degree South Africa – typical forms of suburbanization emerged. Here, more than anywhere else, the construction of the suburbs was instrumental in the 'suburban solution' of capitalist overproduction crises: the massive erection of single-family homes with their associated shopping centres was an ideal platform for the shift of capital from the (glutted) production sector into societal consumption (Walker 1981). Suburbanization created the environment for the consumer society of the post Second World War period. Apart from producing space itself, it offered a close to limitless opportunity to create consumer demand for household gadgets, automobiles, electronics and other commodities during the Fordist era. This historically specific and subsequently defining form of suburbanization is central not just for the shape and the relationships of the suburban and its ways of life worldwide but also for how urban research itself has viewed these processes traditionally. This historical form of suburbanization in America is predominantly tied to three important dynamics:

Home ownership. In Anglo-American societies and more pronouncedly in settler democracies of the 'new world' landed property and especially the suburban home became the symbol of the liberation from European shackles of status and class and of the self determination and cultural autonomy of the individual. The cultural preference for home ownership, supported by state policy and market institutions, is an important marker in the way housing development (and ancillary commercial and industrial development) is structured in space. Suburbanization has been one important articulator of homeownership. The suburb of homes was also a suburb of home*owners*. Alongside the form, there was the private title to the land and to the property that characterized this thrust of suburbanization in Anglo-America. Houses were owned, as was the land, whether those structures were built through a formal, regulated process of subdivision and mass production or a more informal ('unplanned') process of acquisition and sweat equity (Harris 1996). Pitted against 'the lodger evil' and the 'unstable tenant', ownership of the single-family home was considered a societal goal with ancillary benefits of great value. Privacy was seen as the glue of society and the protection of the family all at once. Richard Harris notes: 'In Toronto and across the continent, just about everyone felt that home ownership was a good thing. They believed that it reinforced the independence of the family, encouraging thrift and reducing demands upon the public purse' (Harris 1996: 97). The disciplining effects of home ownership on workers and immigrants were especially emphasized.

While the home and its ownership were factors of social stability and the backbone of a conservative outlook on urbanism overall, the automobile became the connector that provided mobility when needed and always symbolized freedom, perhaps even more than the title to the suburban home. Although, as is well known, suburbanization in its common American form was often the immediate outcome of land speculation driven by rail based infrastructure – with Los Angeles perhaps the classic case – leading among other things to 'streetcar suburbs' (Warner 1978), the private motor car was ultimately the important missing link between the suburban home and the productive and consumptive hubs of the emerging Keynesian-Fordist economy of the mid-twentieth century (Poitras 2011). Taken together, these developments resulted in what Harris (2014) has called 'three suburban stereotypes about North American suburbs': the ideal quest for privacy, conformity with the suburban ideal and the common misgivings among researchers about suburbia (see also Fiedler and Addie 2008). But, of course, even the North American ideal case really had variegation and distinctive periods often neglected by apologists for and critics of suburbia alike (Lang, LeFurgy and Nelson 2006).

Still, homeownership is not a necessary condition as suburbanization occurred in many countries in the twentieth century where there were low ownership rates (Germany, Switzerland) or even no ownership was the rule (the countries of the Eastern Bloc after 1945) but in other countries, the combined state and market supports for home ownership (land and mortgage markets, tax incentives) created the perfect conditions for suburban home ownership as the core of the model of suburbanization we have come to see as typical for the phenomenon (Phillips 2014). In some countries, like the UK, where a longstanding preference for home ownership has existed, high rental rates persisted until Margaret Thatcher's drastic policy changes in the 1980s, when favoured privatization led to a dramatic increase in home ownership and subsequent suburbanization in the form of single-family homes in semi-rural areas. The financial crisis of 2008 also drastically changed the dynamics of home ownership, as some of the countries with the highest rates of increase in both homeownership and suburbanization (Spain, Greece, Ireland, for example) also became hotspots of the real estate malaise that has contributed to the crippling of these economies since (Heeg 2017).

While for generations, in Europe, urban rental housing was considered a desirable distanciation among progressive middle classes from the rural tradition of home ownership often linked to backwardness, parochialism and agricultural roots, suburban home ownership in settler societies, by contrast, was the sign of distanciation from newer waves of immigration that continued to flow to the city centre with its assimilative institutions and job opportunities. Suburban home ownership was the most visible symbol of successful arrival. The condition for it was created by cheap land and energy and the financial institutions that supported the purchase of affordable mass-produced bungalows on green fields at the urban periphery. This was the birth of the 'drive till you qualify' formula of affordable homeownership that is often considered the origin myth of sprawl.

While, following this formula, consecutive waves of ethnic immigrants left the 'slum' of the inner city for both suburban life and home ownership, the racialized housing market in the United States in particular had the effect of keeping African Americans in the 'ghetto' in the urban core. Paradoxically, the so-called 'white flight' to the suburbs (which was more mixed than white to be sure) afforded inner-city African Americans to become homeowners at a faster rate. Some economists have argued that metropolitan Black homeownership rose sharply as Whites moved to the periphery: between 1940 and 1980, the rate went up by twenty-seven per cent (Boustan and Margo 2013).

Industrialization. Suburbanization as an organizing principle of urban space does not just concern the privileged neighbourhoods of the urban middle classes that flee the city but also the land uses by commerce and industry. In this sense, the typical suburb serves the relocation (or new siting) of certain functions from the city. Anti-urban tendencies did not just characterize the residential populations of the immigrant countries (more a produced than an innate preference for sure!) but also the ideologies of industrial pioneers like Henry Ford who deliberately made his famous assembly lines run in the suburbs of

Detroit. Sub-metropolitan industrial districts and metropolitan labour markets, capitalist strategies, sectoral needs and all manner of other factors influence the specific locational mix of the 'multimodal' suburbanizing region overall. To all of this, the location and mix of industry has been critical (Walker and Lewis 2001: 9).

Closely related to industrialization was the massification of housing production itself and the expansion of mobility networks. Baum-Snow, for example, showed that 'innovations to the urban transportation infrastructure played a key role in influencing changes in the spatial distribution of the population in U. S. metropolitan areas between 1950 and 1990' (2007: 776). Massive population losses in inner cities can be credited to the building of interstate highways that both destroyed existing neighbourhoods and enabled moving to the periphery. While cities lost as much as eighteen per cent of their pre-highway population, inner cities might have actually gained eight per cent in the same period had those road transportation networks not been built (Baum-Snow 2007).

Displacement. Suburbanization has historically been tied to displacement. The classical example for this dynamic stems from the nineteenth century when Baron Haussmann, on behalf of King Louis Bonaparte, fundamentally modernized medieval Paris: 'He gutted Paris according to plan, deported the proletariat to the periphery of the city, simultaneously creating the suburb and the habitat, the gentrification, depopulation and decay of the center' (Lefebvre 2003: 109).

The central figure of thought here, the push of the proletariat outward, has become a staple in what Neil Smith (2002) has called 'gentrification as a global strategy'. Indeed, everywhere in the world, in places as different as Istanbul, Paris, Mumbai, Shanghai, Johannesburg, Toronto, Mexico or Sydney, poorer populations are pushed out of regenerated city centres and into new built or restructured inner and outer peripheries.

The notion of an escape from the city has a particular significance in the historiography of suburbanization. It denotes a foundational tendency that is captured well by P. D. Smith: 'The suburbs offered a safe haven, a bourgeois utopia whose semi-rural location made possible a relaxed, outdoor lifestyle which affirming values deemed central to American society, such as the sanctity of the family and property ownership' (2012: 145). The American case displays a historical specificity in this context which overdetermines the debate about suburbanization until the present time. The so-called flight to the suburbs has been strongly overlaid with racializing or even racist tones. Suburbanization has often been discussed in connection with what has been called 'white flight', that is the deliberate and planned distanciation of white people of all social classes from African Americans who became increasingly concentrated (and segregated) in the inner city of the United States. While this particular expression of white privilege in the context of suburbanization is specific to the United States, suburbanization as a socio-spatial distanciation strategy is a universal phenomenon. Especially today, in the age of privatized suburban developments, the so-called 'gated communities', the social walling-off in the space of the private home at the periphery is more and more a general experience and common form of expression of the suburban reality.

EXPLAINING SUBURBANIZATION CRITICALLY

Critical urban research has been traditionally sceptical towards suburbanization, especially among Marxist and Feminist scholars. The common (and generally warranted) view that suburbs are a bulwark of the political and social reaction (see for example Harvey 2013) consequently led to critical research perennially ignoring and caricaturizing suburbanization. Interestingly, however, the slogan 'The Right to the City', minted by Lefebvre at the end of the 1960s, was formulated as a result of the urban deficits that were felt in the periphery of Paris. In this respect, the suburbs opened up key questions of the survival of Fordist capitalism, when the crisis of the 'never ending prosperity' of the post war years hit the cities in the 1970s.

The markedly different experiences of European suburbanization did not just include a movement of the middle classes into the single-family home subdivisions at the city's edge but also a massive shift of the working classes to large housing estates on the 'green' periphery. This occurred predominantly in the Fordist core countries in the West, Great Britain, France and West Germany but at an even larger scale in the countries of 'real existing socialism', where tens of millions of housing units (and large numbers of factories) were built in peripheral urban extensions. Between 1950 and 1990, in the European socialist countries 'the urban population of the region almost doubled, increasing its share from 38.3 to 66.5 percent' (Stanilov and Sykora 2014b: 3–4). While the Eastern bloc countries lacked the de-centralized, low density suburbanization, its peripheral housing estates were 'rarely located at a distance from the compactly built-up urban areas. They were planned as an integral part of the socialist city, functionally integrated with industrial zones and service nodes through public mass transit infrastructure' (Stanilov and Sykora 2014b: 6). In the second half of the twentieth century, in most Eastern European cities more than half of the population lived in pre-fabricated, large-scale housing estates (Hirt 2012: 35), often on the sharply defined urban periphery in extensive, dense but spacious, sub/urban extensions such as New Belgrade or Prague's South City (Logan forthcoming). Different in many ways from their Western European counterparts (Hirt 2012: 34–6; Stanilov and Sykora 2014a), they nonetheless represented a massive expansion of urban form and function beyond the classical footprint of the European city.

Among the buzzwords of the era were new towns, satellite and 'trabant' settlements that were to give the sprawling metropolitan regions of the Fordist (and socialist) period structure and form. They built on prior traditions of garden cities and greenbelts such as the affordable housing programme *Das neue Frankfurt* of the 1920s when housing estates were built like a pearl necklace around the German city. In Vienna, the city edge was accentuated with the Karl-Marx-Hof. In Zurich, mass rental housing in midrise bars and towers, nondescript office factories and office parks grew (and still do) between Schwamendingen, Oerlikon and the airport to absorb overflow capacity from the city and to receive international migrants and refugees who have moved to the city in growing numbers.

In contrast to the mostly privatized suburbanization in North America (albeit subsidized through tax and other incentives by the state), suburban housing during high Fordism in Europe was generally produced by public or co-op development and management corporations. It has got to be kept in mind, though, that the institution of the savings and loan industry in Germany made a significant ideological and material contribution to the proliferation of the single-family home suburban form. This played strongly to the philistine tradition in Germany to which property ownership probably had a similar significance as in the United States.

In the West as much as in the East of Europe these estates became the focal points of a new urban crisis at the beginning of the twenty-first century, as the normative attention of the planners and developers started to cultivate buzzwords such as re-urbanization, New Urbanism and the creative city. Some countries, such as Canada for example, have begun to develop mixed forms of suburbanization where North American style single-family homes are being rolled over fertile farmland but European style high-rise neighbourhoods are built on the periphery (Charmes and Keil, 2015).

STRUCTURAL CRISIS AND THE ENFORCEMENT OF SUBURBANIZATION AS THE GENERAL FORM OF URBANIZATION

After the 1970s, Western capitalism reacted to its structural crisis with a fundamental re-evaluation of spatial and urban development. In the countries of the former Soviet zone of influence, a process of turbo privatization was released after 1989 which sidelines the state as an actor in the urban development process (East Germany remained the exception with its richly funded *Stadtumbau Ost* urban regeneration programme). The liberalization and deregulation of urban planning is leading to sprawl and fraying beyond traditional urban cores and peripheral high-rise estates (Hirt 2012; Stanilov and Sykora 2014a; Szirmai 2011). In the countries of the Global South, urbanization exploded predominantly through the country-to-city migration since the middle of the twentieth century, first in Latin America, more recently across Asia and in Africa. The majority of the newcomers move to informal urban peripheries, often in auto-constructed squatter settlements (Caldeira 2016). The destinations of this rural to urban migration have been called *Arrival Cities* (Saunders 2011). In the metropolitan regions of Turkey, China and India, large-scale high-rise neighbourhoods surround the urban centres that have become the almost exclusive residential zones of the middle classes. There and in Brazil, South Africa and other parts of the developing world, synthetic mass housing, mostly in high-rises, is being plunked onto greenfield sites on the peripheries of major cities, sometimes even in the sea or on reclaimed land as is the case in places as different as Amsterdam and Hong Kong (although there is persistent direct contiguity between the richest and poorest zones as, for example, in the favelas of Rio de Janeiro). A large part of this primary peri-urbanization occurs in gated communities (Bloch 2015; Gururani and Kose 2015; Mabin 2013).

Three intertwined processes have enabled and furthered the increased suburbanization and have made suburbia the dreamscape of the current epoch:

Postfordism. In the immediate period after the Fordist crisis, much attention was paid to the rapid shift in socio-spatial structures that could be observed universally. A flexibilization of spatial forms replaced classical functional segregations that had dominated urban development after the Charter of Athens in the 1930s. The new normative ideal became mixed use as both urban centres and peripheries started to change. More important than the normative ideal was the tendencies among core post-Fordist industries to settle in business zones on the peripheries, where they did not have to face contaminated sites, where land was cheap, where no traditional labour union organization existed and where, especially, the relevant infrastructures for the new just-in-time economy existed. The new exurbs were much better placed to service this economy than the older inner cities especially when developed around major airports (Addie 2014).

Globalization. While the globalization debate in critical urban research has concentrated primarily on Global Cities and their centres, the urban peripheries became the actual stages for the performance of global economy and culture. The global city literally went up the country, away from the command centres of the financial industry and advanced producer services, but towards other important nodes of the internationalized global economies. In almost all metropolitan regions in that period, the exurban airports and their surrounding regions have become centres of economic activity and relay stations of the globalized world. In addition, in many urban regions the suburbs have become magnets for immigration, so-called ethnoburbs.

Neoliberalization. Simultaneously with the development towards a post-Fordist economy and with globalization and deeply entwined with them, capitalism worldwide is undergoing processes of neoliberalization. In this period which places the privatization of economic decision making and responsibilities over collective solutions, suburbanization proves to be an ideal field for a comprehensive restructuring of social and spatial relationships. The product has been an especially 'vulgar' form of urban development, with ostensibly displayed wealth that expresses itself in supersized homes and strongly-guarded security zones (Knox 2008; Peck 2015a).

Suburban cityscapes are now so ubiquitous that they have led to a re-evaluation of urban form and process overall. An important turning point in the perspective on the European city in particular was set by Tom Sieverts's 1997 publication *Zwischenstadt*, which appeared in English in 2003 and threw into sharp relief the multiply dissected, no-longer-central structure of suburbanized landscapes. Internationally, the new metropolitan regions have been discussed under the keyword 'post-suburbanization' – a term mostly trying to denote that historical suburbs have now entered themselves into a complex renewal and urbanization process (Charmes and Keil 2015; Phelps and Wu 2011; Phelps 2015).

These internally imbricated processes open the opportunity for a fundamental discussion of the urban-regional governance problematique and of the politics of the suburbs. Suburbanization itself had originally induced the necessity for a regionally scaled politics as existing centres and new suburbs had to calibrate their mutual relationships. Suburbanization also enables a new type of regional politics that has reoriented itself in a world that is characterized by 'horizontal strategies of surveillance, dispersal and consumption' (Quinby 2011: 139). This politics, however, is no longer just about maintenance of the status quo, but it affords new emancipatory political dispositifs beyond the traditional hierarchies and confinements of the urban territory.

Finally, suburbanization permits insight into real existing processes of urbanization in the twenty-first century and hence into the emergence of 'urban society' (Lefebvre 2003). Politics consequently does not necessarily come from the centre. It increasingly comes from the desire of the periphery and the maelstrom of the post-suburban city. Suburbanization presents itself as a broad spectrum of metropolitan problematiques: long commute times, insufficient public services, violence, underemployment and unemployment, multicultural frictions and solidarities, school and educational policy, environmental and cultural questions and many more. In its diversity, suburbanization, ultimately, shows itself to be the magnifying glass through which we can get a better glimpse of the 'urban century' than through the tunnel vision onto the monocultures of the gentrified inner cities.

Istanbul, Turkey

3 Suburban Theory

Having explained some of the main tenets of suburbanization and suburbanisms in their historical and geographical diversity, we are now moving to suburbanization and suburbanisms as objects of critical theory. Urban theory has traditionally been an exercise of understanding the city from the inside out. This chapter will deliberately upend this tradition and attempt to challenge, reverse and propose to eventually abolish the centralist bias in urban theory and introduce a new way of looking at what we call 'suburban' in the context of urbanization processes overall.

Roughly two decades into the urban century and two generations after the realization that we were entering the state of planetary urbanization, mainstream observers and specialized students of urban matters alike begin to fathom the complex nature of city-building ahead. The story that is coming to the fore now is that this urban society will bear no or little resemblance to the old imagery of urbanity that we have carried with us since the nineteenth century when Paris, London, Chicago or Manchester gave us urban models to work from. If anything, now, we have arrived in the suburban century or, better, in a post-suburban century where peripheral growth is the norm but that growth is vastly different from the kind of suburbanization we encountered in the second half of the twentieth century in North America, Australia and Europe. This is the age of the urban periphery.

Much of what we will witness over the twenty-first century in terms of human settlement will take place in those areas in the world that have traditionally not been the location for the building of cities and the building of theories. This is perhaps the most important point. Non-central worlding practices now include strategic suburbanizations as part of a multilogue of inter-referencing in the Global South itself (Roy 2011, 2015).

Instead of idealizing an urban theory that is built on urban centrality and density (as is the case in much post-Jane-Jacobs urban thinking) and in contrast to facile attempts to equate suburban life with human freedom (as is the case with American libertarians), I subscribe to basing our urban theorizing on the basis of the real existing post-suburbanism in which most people will spend their time, living, working and playing in this twenty-first century.

Interestingly, critical urbanists have left much of the thinking if not theorizing on the new sub/urban forms to those who are not urban thinkers. In a write up on biologist Karen Seto's work in the science journalism outlet *Undark*, we learn that suburb is a loaded term:

> "'Suburb' doesn't really make sense in Bhutan or Nepal or even India," Seto says. Our typical words to describe urban forms – city, mega-city, suburb, exurb, slum – don't capture the diversity of new urban growth. "What's very clear is that we need a language and a finer-grain differentiation of different types of urban life and urban ecosystems," she says. (Humphreys 2016)

Seto's scientific approach allows us to measure the sub/urban expansion. She can see the problem but lacks the toolbox needed for its solution. This is where urban studies in general and urban theory in particular are desperately needed.

It is, thus, time to argue for an intervention into urban theory on the basis of the suburban explosion. Today, we are facing new realities of globalized urbanization, where central city and fringe are remixed. In this period of the post-suburban planet, new forms blend together on the periphery; the result has been a different suburbia and a different city. In simple terms and following the definition of historian Jon Teaford (2011: 33), global post-suburbanization refers to myriad forms of 'suburbanization carried to the extreme, the end product of two centuries of continuous deconcentration of metropolitan population' (and one might add of economic activities). In building urban theory from the outside in, we take our cue, as we have before in this book, from Henri Lefebvre who argued: 'The explosion of the historic city was precisely the occasion for finding a larger theory of the city and not a pretext for abandoning the problem' (Lefebvre and Ross 2015). Importantly for Lefebvre, as Walks (2013: 7) has argued, 'the suburban is conceptually an extension of urbanism' but at the same time since suburbanization has appeared on the scene as a massive phenomenon, it becomes the basis for a new and more comprehensive theory of the city and ultimately society.

Accordingly, as Klaus Ronneberger has put forth in a careful reading of Lefebvre's work on centrality, Lefebvre understood the urbanization process ultimately as polymorphous with fragmented relations in a disparate urban fabric that is characterized simultaneously by processes of concentration and deconcentration and centres and peripheries of variable dimensions (Ronneberger 2015: 26). Theoretically, both for Ronneberger and Lefebvre, the city is ideology and the urban is a space of contradictions that longs for theoretical explanation and political intervention. This urban space includes both the imploding centralities of the core and the exploding vectors of urbanization that surround it (Keil and Ronneberger 1994). Yet it remains necessary as we consolidate the dichotomous fragmentation of non-related categories into dialectical syntheses, to understand the merit of individual processes that make up the overall dynamic of the planet's move towards complete urbanization. In this dynamic, distinct processes of suburbanization contribute to the overall push towards an urbanized world. Those processes can be understood in the foundational categories of governance, land and infrastructure and can be studied empirically in those contexts but they also need to be seen as important elements from which to draw theory. Returning to Jenny Robinson's appeal that '[c]onsideration needs to be given to the difference [that] the diversity of cities makes to theory' (2002: 549), then, we can note that consideration needs to be given to the difference that the diversity of suburbs and suburbanization more generally makes to theory.

FORWARD INTO CRITICAL SUBURBAN THEORY

How have critical urban theory and research benefitted from prioritizing the peripheral and how has a dialectical approach, in turn, helped transcend the urban and suburban divide in theory and practice? Around the end of the first decade of the twenty-first century, a group of scholars centred around the City Institute at York University deliberately made suburbanization and suburbanisms the focus of a large collective research initiative which attempted to apply critical theoretical thinking to the process of suburbanization and the everyday lives lived in the urban periphery. While it is to be expected that such a project today would be multi-disciplinary and multi-method, it was specific to this project and new to the overall endeavour of critical suburban studies that the approach would be global. The intended globality was going to be established through four related processes: (1) the team of ultimately fifty associated scholars was to be roughly representative of some of the major existing literatures and ongoing research efforts worldwide of peripheral urbanization; (2) while it would be impossible to cover the actual global multitude of cases of suburbanization and suburbanisms, we made a sincere effort to be as geographically inclusive as possible; (3) the thematic areas of the research were chosen with an eye to relevance in suburbanization processes around the globe and (4) and most importantly, we pledged not to make this an exercise in applying preordained 'Western' theory to multiple case-studies elsewhere. Instead, the inclusion of urbanization in the Global South (and other subaltern real and imagined regions such as the postsocialist countries) in the debate on global suburbanism(s) is not a mere addition of more empirical cases to an existing script of peripheral expansion.

It was a precondition of our research not to assume convergence and conceptual universality. There is no single global suburbanism. The project conceptually and ethically defied the notion of the North American experience as a model for all suburbanization; and it made every effort to be empirically inclusive. We continue to examine the dazzling and puzzling diversity of the suburban process, form and function (Keil and Hamel 2015: 357). This adds up to a de-centring of the geopolitics of the urban narrative. New processes originating from the margins and peripheries defy the traditional dependencies of outsides from insides, suburbs from cities and expand our understanding of the dialectic of the urban process. Finally, suburban studies allows for a more radical theoretical inversion of urban theory: it is from the emerging geographies of non-European and non-American (sub)urbanity that the architectures of urban theory await rebuilding.

REBUILDING SUB/URBAN THEORY

One way in which scholars and practitioners have begun this journey is to work deliberately on the discursive constructions that undergird processes of suburbanization and suburban ways of life. This remains a challenge. For now, we have settled for hybrid terminologies. Some of us speak of the peri-urban, others have subscribed to more mixed concepts such as *Zwischenstadt* (inbetween city) (with reference to work by Tom Sieverts (2003)) or post-suburbanization (as pioneered among others by Nick Phelps and Fulong Wu 2011). At the tail end of those iterative and inductive travels through suburbia, scholarship will at one point come across new languages that will adequately describe and allow us to analyse, the sub/urban world in which we now live. Suburb, suburbia, suburbanization, suburbanism are themselves marked terms as they conjure images of middle-class, white, single-family home peripheral settlement in North America. Richard Harris and Charlotte Vorms (2017) are tackling this challenge in an edited volume that exposes and explicates the Babylonian plurality in naming settlement away from the core in countries around the world. They have catalogued the wide-ranging variety of the suburban nomenclature. Difficulties inherent in naming the global suburb (or the processes that produce it and the ways of life that it enables) result from that very variety. With English as the lingua franca and default language of sub/urban research, bringing the actual literatures of different national discourses into the mix is central. Engagement with many traditions of suburban research and urban theory in French, Spanish, Brazilian/Portuguese, German, Chinese language traditions, as well as with Eastern and Western European, Indian, African, Australian and Latin American ways of thinking and doing research is now *de rigueur*.

But it is not just diverse and differential lexica that are at issue here, it is the very understanding of urban life as it is constituted through historical state and market governance. As Ananya Roy (2016), for example, has pointed out, what is urban is a matter of differentiated historical geographies and she challenges us to 'pinpoint the conjunctures at which the urban is made and unmade, often in highly uneven fashion across national and global territories' where often 'urban' or I might add 'suburban' is a governmental category (in Roy's case she refers to urban and rural). But the push really needs to go beyond the naming and the categorizing alone, both important endeavours in themselves. The push needs to be for writing urban theory from the outside in.

The variegation in concepts and real forms and processes, then, will engender direct and calibrated engagement with theoretical 'why' – questions such as 'why suburbs now' and 'why suburbs in this form and shape'? Those questions will be met by a number of explanations that tell us more clearly what the world is like that we live in. The grammar of the suburbanized landscape, which is rolled out across the globe, is as good an indicator of where and who we are, as Manchester was to Friedrich Engels in his famous reflections in the middle of the nineteenth century. This works in much the same way as Chicago represented the screen for the researchers in the School of the same name in the 1920s and 1930s or Los Angeles was the blueprint, for a while, for the Los Angeles School of urbanists who predicted a quite different sub/urban future. The causalities, i. e. the 'whys' are beginning to reveal themselves through comparative, multi-site, multi-disciplinary, multi-method empirical and conceptual research and will lead to theoretical and conceptual insights that will guide further research on the topic.

URBAN THEORY AFTER THE SUBURBAN EXPLOSION

Just as the 'worlding' of formerly peripheral areas in the Global South has been observed in recent analyses, the presence of the global suburb in the intellectual domain of the global city can no longer be ignored (Gururani 2013; Herzog 2015; Mabin 2013; Maher 2004; McGee 2013; Ortiga 2016; Robinson 2006; Roy 2009, 2015; Roy and Ong 2011). As noted previously, critical urban research has been traditionally sceptical towards suburbanization. In as much as 'the essence of the modern suburb is physical, social and spatial separation' (King 2004: 99) its analysis may not have revealed much in general terms about the city as a theoretical object.

This may be entirely different today. The global suburb is – perhaps – less about dichotomous splits and more about multifarious connectivity, although often 'the social relations of global suburbs reinforce many of the same inequalities as in the traditionally segregated city' (Maher 2004: 804). They just do so differently. Suburbanization used to be a process of distance-making: classes, ethnicities, race, and so on (Ibid.). Now there is a partial reversal going on: Suburbanization turns into a process of adventurous mixing and reshuffling of the urban, while the city (as much as it still exists as a recognizable unit) becomes the rarefied monoculture of condominium-dwelling creatives that operate in safe, sterile, predictable environments.

This is the shift from Castells's wild city to the 'wild suburb', from Lefebvre's 'habitat' to 'the world's city' (Castells 1976; Lefebvre 1996). Global suburbs, globurbs and ethnoburbs are the new assemblages of the global that surround our cities and, as they beg for explanation, yield a host of new insights about urbanization overall (King 2004). The global suburb explodes and reassembles the very categories we used to associate with the single-family home behind the picket fence: home ownership, industrialization and displacement. The global suburb is somewhat of a pendant to the global city, although it is not clear whether it carries the same cosmopolitan caché or if it is likely going to 'generate anxieties about both physical and social boundaries' (Maher 2004: 782). In any case, there is evidence that global suburbs, with their immigrant-based service economies, will involve significant social stratification as well as elaborate and multiple social distinctions (Maher 2004).

Suburbia allows a view into novel property relationships between community and land, urbanization beyond industrialization and enclaving beyond displacement. At one end of this new spectrum of suburban form and life, we find the opportunity to redefine suburban everyday life altogether (Drummond and Labbé 2013). In this context, one would expect to find modes of organizing everyday life in ways that neither find themselves locked in an unchanging state of stasis (as the classical commuter suburbs of the twentieth century were often caricaturized) nor imagined to be catching up to a certain ideal of urbanity that draws its inspiration variably from the medieval Italian city, Barcelona, Copenhagen and New York. Much of this development depends on new technologies and the rapidly evolving complexities of peripheral communities that seek their organization beyond the societal model of the Anglo-Saxon tradition. At the other, we find the dialectics of territorial control, the martial state, social neglect and private authoritarianism;

between new types of poverty and the wealth of gated communities. All across the global suburban multiverse, we find a horizontal landscape of consumption (King 2004; Quinby 2011).

Suburbanization is no longer a simple, concentric extension of existing urban morphologies. 'Extended urbanization' presents a number of conundrums (Lefebvre 2003). In the Global South, where up to two thirds of regional populations live in suburban areas, the morphologies, composition and even concepts of suburban development abound (McGee 2013; McGee 2015). In future, those suburban fields will house the majority of those yet to move to cities for the first time (Swilling 2016). Roy (2015: 342) detects 'a patchwork of valorized and devalorized spaces that constitute a volatile frontier of accumulation, capitalist expansion, gentrification and displacement'. And 'peri-urban areas' are often high density and include a wide and dynamic variety of forms and land uses (Gururani 2013; Mabin 2013). The dynamics of emerging and existing suburbanization in the Global South add to a kaleidoscopic global suburban landscape. There is plenty of idiosyncrasy and endogenic activity in the myriad suburban forms that are now emerging; there is also much blurring among and between the morphologies, lifestyles and infrastructural technologies in different world regions.

In an ubiquitously suburbanizing world, we can differentiate between extended urbanization – that which occurs in highly urbanized but demographically stagnating industrialized regions and primary urbanization, suburban formation which occurs in high-tech or resource areas (oil-burbs, etc.). Primary urbanization forms include gated and otherwise access-restricted enclaves of privilege; slums of the displaced and squatter settlements of the hopeful; sprawling single-family home subdivisions; hyper-dense tower neighbourhoods; new developments pushing into ecologically sensitive areas; and existing, though sometimes abandoned, undervalued inner suburbs that await renewal and regeneration. Add to that residential districts of variable density and industrial or commercial districts where service infrastructures and non-desirable land uses are concentrated.

In this process, historically bland and clear-cut segregation(s) in land use, socio-economic make-up and socio-natural relationships associated with suburban form and life begin to break down into more complex variations. This globalizing suburbia, then, presents itself as a historically evolving human geography: the predominant human habitat is less urban than most observers of the urban century assume and much less rural than where people lived one hundred years ago. It is time to face the emergence of a post-suburban planet where existing and new forms of peripheral urbanization interlace in a complex pattern of urbanity.

What urban theory increasingly learns from suburbanization is how the making of peripheries contributes to the building of new centralities. A simplistic view that sees one territorial centre in contradistinction to undefined (and implicitly less important) peripheries now gives way to multi-scaled centre-periphery assemblages with multiple centralities and peripheralities that sustain urban regions (Keil and Addie 2015). Global suburbanization has imposed upon us the imperative of challenging the models from which it ostensibly emerged and forced us to change the ways we study suburbs in a world of complete urbanization. Lefebvre's defiant plea in response to Guy Debord and his associates among the Situationists of the 1960s, not to 'abandon' the city and urbanism as arenas and pathways of societal change was prescient. His intellectual provocation of sticking to 'the urban' as a source of theoretical insight, revolutionary potential and societal transformation, just as 'the city' dissolved into the urban, proved to be of lasting significance (Lefebvre and Ross 2015: 50).

The call for a renewed urban theory is once again before us today. The 'explosion' was merely the starting gun of a much more far-reaching process of complete urbanization that turns the categories of geographical centre and periphery upside down and creates new relationships among various parts of urban regions. It points to a more profound entanglement of constellations than the original astrophysical metaphor of the explosion would suggest.

We have, for example, now lived for two generations with Lefebvre's 'little houses' and the towers in the park on the urban periphery. We have seen libraries of writings on communities of tree-lined streets with white picket fences; work that is ostensibly filled with the small and large pathologies of modern life. The social critique, the political critique and now the environmental critique have not been kind to the single-family home suburb. Yet it persists. We have been through waves of real and intellectual demolition of the modern dream of the high-rise periphery, while China, Singapore, Vietnam and Turkey, as well as many other countries, continue to produce them at breakneck speed. And we have seen reams of publications on the future of a 'planet of slums' (Davis 2006). Suburbanization today enables insight into the real, existing processes of urbanization in the twenty-first century and hence into the emergence of 'urban society'.

THE INCOMPARABLE SUBURB

Global suburbanism then becomes a programmatic intervention into the theoretical and empirical investigation of the massive worldwide proliferation of suburbanization in the twenty-first century. Rather than focusing only on (sub)urban form, this re-evaluation of global suburbanization also takes into account the political economy of urbanization, the governance of peripheral space and the infrastructures supporting suburbanization globally today.

There is no teleology of suburbanization. Not all suburbs grow up to be cities. There is no direct line to urban paradise from the periphery. Some suburbs go straight to post-suburban hell. Others continue to define the horizon of possibility for aspiring, urbanizing masses from the rural countryside and from other parts of the urban world. Today's suburban areas are born into a post-suburban world. It is inconceivable today that suburbia could be left in some timeless state of never changing bliss (as some homeowner associations might want it to be). The opposing extreme is equally implausible: suburbia as a juvenile version of a future city (as some believers in the transformative powers of urbanism have it). We now know that either option is unlikely. We must also abandon the dichotomy of single-family home subdivision on one end of our imaginations and the tower dominated 'badlands' on the other. The in-between cities we find in the post-suburban landscapes of our 'globurban' reality will have traces of both, but rarely monocultures of either. This ultimately means that urban theory that is prompted by global suburbanization is necessarily comparative in nature (Nijman 2015; Peck 2015a, b). Only, now the comparison is increasingly losing North American suburbanization as its assumed yardstick. We are now in a field with fast moving referents that belie many of the old assumptions on suburbs and urban theory. It is time to bid the (conceptual) white picket fence farewell.

ENTERING THE SUB/URBAN CENTURY OF THEORY

Many scholars have subscribed to a theory, first pioneered by Henri Lefebvre in the 1960s, that there is a 'virtual object' we can call 'urban society' in which even those areas that are not obviously urban in form are part of a process of 'complete urbanization' (Lefebvre 2003). For Lefebvre, this emergence of urban society is linked to a series of dialectical processes of implosion and explosion. Not enough attention in this intellectual tradition has been paid to the constituting properties of the explosive antithesis to the implosive thesis. If we assume, as we have in this book, that most urbanization today is suburbanization and urbanism as a way of life has increasingly been overshadowed by multiple suburbanisms as ways of life, we may need to reassess the implosion-explosion dialectics: this nuclear-physical notion leaves the point of origin in the centre intact. Today, the generalized horizontality of global suburbanisms forces us to rethink, at least to a degree, the core beliefs of urban studies: we need to abandon the centripetal-centrifugal orientations of urban theory (that allow explosions only as a movement away from an imagined centre) and take Lefebvre's virtual object seriously as a generalized condition of post-agrarian and post-industrial possibility: urban society as a set of multiple centralities that are neither geographically nor functionally linked to a pre-existing traditional core.

Through this theoretical move, we establish suburbanization as a process removed from the mono-directional shift away from a given, unquestioned centre. In the past, there was reason to believe in the existence of two extremes – and this is borne out through most settlement processes we have experienced: that suburbanization would either be an early, incipient process of urbanization or lead to the relegation of suburbia as a forever lesser valued form of collective settlement. In the first instance, suburbs would be considered to mature as, for example, most so-called early twentieth-century streetcar suburbs did. They became part of Lefebvre's *tissu urbain*, ultimately indistinguishable from their original urban environment. In the latter instance, those suburban areas would be forever held in a condition of heightened pathology, a chronic state of deficit – of culture, wealth, urbanity or anything else on a register of values for which the real and imagined 'city' remained the yardstick. *Successful* suburbanization in this view would be considered delivering the suburbs into the city, making them disappear into their Other, the traditional city. *Failed* suburbanization was that process that left suburban neighbourhoods behind, set them apart, made them unfit for urban living, devoid of social and morphological mix and integration, a non-place for constructive political power and – most importantly in an age of climate change – utterly unsustainable. Remarkably, of course, these pathologies could be equally stated for the very rich, gated, privatized communities as for the poorest, ghettoized, racialized fringes around cities as found in much of Europe and Canada.

In this book, I present a different view. Suburbanization is now a pervasive, open-ended, negotiated process of settlement in which neither success nor failure in the above register may be relevant categories of analysis and understanding. Instead, those suburbs that will now house and employ most of the globe's population will forever be different from the cities of the past, neither melt healthily into those, nor stand apart in pathological manner. In following chapters on diversity, ecology and politics, we will explore the processes of the suburban revolution in this sense further.

THE SUBURBAN PROSPECT: OBSERVATIONS ON GLOBAL SUBURBANIZATION

In line with the argument put forth in this chapter, I am therefore assuming that the majority of the new urban forms and processes brought on by this tremendous wave of urbanity will be suburban in nature. There will be some massing of built form and concentration of urban process but given the uneven distribution (Africa and China) and specific path dependencies of those urbanities, the majority of urban settlement will not just be geographically peri-urban, outside of the current perimeter of urban areas in those areas, but also – at least in Africa – mostly low to middle density in nature. Even in China's massive tower neighbourhoods that are pushed outward concentrically, ringroad after ringroad (Fleischer 2010), the majority of new and existing urbanites will live in suburban constellations (Keil 2013; Swilling 2016).

The rise of the suburb(s) is a global phenomenon but those peri-urban extensions to existing urban form and structure are locally built and re-built. There is a large diversity in process and outcomes of suburbanization. This diversity in form and structure is reflected in diversity of concepts in a 'world of suburbs' (Harris 2010). Suburbs also must not be considered an organic product of aggregate consumer choice but the composite outcome of planned and regulated interactions of public and private actors in very different systems of land economies and governance (Hamel and Keil 2015).

This conjuncture also denotes a break in the history of urbanization: from increasingly dense and mostly industrial Manchester or Chicago, or finance capitalist New York, London or Tokyo, we gravitate to suburban expansion everywhere. As this 'exploding' urbanization (Lefebvre's 2003) proceeds, 'imploding' centralizations will continue to take place, leading to a diverse global (sub)urban landscape. As we shall discuss in more detail in Chapter 8 on suburban political ecologies, these explosions and implosions are themselves planetary in their reach, as global metabolic processes constitute and rely on the ubiquitous suburb.

Suburbanization appears now as original and constitutive, not a derivative element of urbanization. To paraphrase Harvey's (2007) observation, based on Mike Davis's 'planet of slums', we can speak of a 'planet of suburbs'. This means that the epistemic lens through which we can know the urban changes in significant ways. There is a growing need to put Lefebvre (2003) – who had bemoaned the lack of urbanity in the periphery in the 1960s – from his head onto his feet: now the periphery is not just about deficit of urbanity anymore – albeit there are some that are linked to the impoverishment of older, 'inner ring' suburbs in North America and Europe, for example (Charmes and Keil 2015). The geographical periphery, the outskirts, the peri-urban claim new kinds of centralities.

In this new landscape, spatial peripheralization goes along with social marginalization and/or the sequestration of privileges both in classical gated communities and in newer forms of segregation, such as condominium complexes even in suburban hubs. Suburbanization as a general phenomenon now has to be understood on a continuum that stretches from crisis management (Keynesianism), *crisis switching* (Harvey 1982), or in other words, *'the suburban solution'* (Walker 1981), to being recognized as a cause of crisis (subprime crisis, etc.). While most of the talk by well-known urbanists is about the creative cores of the urban revolution, suburbanization is now an increasingly dominant arena for the socio-material process of the production of urban space.

We need to ask now whether, once all is suburban, it still makes sense to speak about suburbanization as a process, suburbs as a place and suburbanism as a distinct way of life. Or should we begin talking about a post-suburban world? These questions are not rhetorical and deserve an honest and productive answer. There are fundamentally two ways to respond. First, in this dramatic period of peripheral urbanization that spans the world, we are seeing a ramping up in scale and intensity of the process of suburbanization, daily expansions of actual suburban areas and a worldwide spread of distinctly suburban forms of everyday life (Drummond and Labbé 2013). In that sense, at least, we can continue to speak of suburbanization.

The second answer is a bit more evasive but still true and useful as it shifts our perspective and course of action in urban theory building to a process of pervasive post-suburbanization. Nick Phelps and Fulong Wu have argued that post-suburbia is a 'composite picture' because of its global manifestations, divergences and mixing of land uses, less predictable geographic forms, new politics, new work-residence relations and discordant land use (Wu and Phelps 2008). We now see a more reflexive process that consists of both the retrofitting of existing suburbs and the continuing emergence of 'original' suburbanization: we can call this post-suburbanization as it points beyond the traditional form of linear peripheral development (Charmes and Keil 2015).

In parallel with post-suburbanization, we can also observe a shift in the *meaning* of the peripheral suburban form. Suburbanization, long seen as the material process of Keynesian-Fordist economics with its virtuous circles of programmed mass consumption, has now become a prime terrain of neoliberalization. Paul Knox (2008) has noted the emergence of a particularly vulgar form of capitalist spatial fix which, over the past two decades, has given neoliberal US-society a specific (sub) urban form of ostentatious consumption of space, nature and resources through large lot monster homes, often in secluded treed domains. Jamie Peck (2015a) has pictured suburbs in the US as privileged sites for the roll out of actually existing forms of neoliberal governance. These observations fit neatly with the notion that the urban periphery has been the 'dream space' of post-Fordist restructuring. The emergence of clusters of high tech or other production, logistics and commercial complexes in wider urban regions is now visible all across the globe, but was first noted in places like Silicon Valley or Orange County in Southern California (Soja 1996). Similarly, the global city economy, tightly linked to the global post-Keynesian evolution, is expanding into the region in search for space and function. This has been the trademark of that period of urbanization.

Suburbs are now a global phenomenon, not just the specifically American spatial fix. Ethnoburban developments are one of many forms of that development. The competence and institutional arrangements of integrating and differentiating populations normally identified with the inner city has now moved to the suburbs (Saunders 2011). But these often informal peripheries don't follow a pre-trodden path towards some more formal, fixed and central. While they are 'cities-in-waiting' as they gel into something more permanent, this permanence will not be anything like the outcome of industrialization-based urbanization in its twentieth-century Western form (McGuirk 2014). Yet, the global periphery is now not seen as the periphery of globalization anymore since its 'worlding' has been part of the overall shift of the geographies of theory towards a post-Western historical geography of urbanization (Roy 2009; 2015).

It is difficult for urban studies to concede that most urbanization is now suburbanization which creates a theoretical conundrum. Through textbooks and in introductory university courses in urban studies, we have consistently fostered the notion that 'normal' urbanization tends to produce centrality and even the sprawling metropolitan areas of the twentieth century have been imagined as agglomerations first and foremost. The suburbs have been largely written into a theoretical space of second-order importance. They were either considered on their way of becoming urban eventually; or worse, unchangeable and timeless in their geographical and semantic location. Suburbanization rebels against us, urban intellectuals, and our sense of self as we cannot imagine the suburban to be part of our personal lives or worthy of serious investigation: they lack the centrality from where meaningful discourse springs. They are the colony to the centre from where we usually construct our narratives and theorizations. The suburbs in the urban imagination appear largely as *terra incognita*: An unknown world, a colonial space. In the worst versions of this mental displacement in urban theory, they become the equivalent of Tolkien's 'Mordor': the place where the horrors of industrial and urban society are played out in the peripheries of the mind (Jeffries 2014). The suburbs have been made extraordinary and pathological, always distant from but ideally on their way to being a normalized urbanism of centrality. They are not part of a dominant geography of theory (Robinson 2006; Roy 2009).

CONCLUSION: THE NEED FOR A REORIENTATION

In the light of this theoretical unease with all things suburban among much critical and mainstream urban theory, there is a dire need to abandon historically privileged spots for observing urbanization. This includes both the privilege of the urban centre and the privilege of the Global North. Many downtowns are now disneyfied, predictable and uniform. They are often gentrified monocultures, filled with green infrastructures (New York's High Line), privileged spaces of recreation and highly stratified economic machines with creative workers in charge and provided for by a precarious service class. Many suburbs, on the other hand, are raw, unpredictable and diverse. They are zones of vast differentiation in built environment (Charmes and Keil 2015), mobility access (Keil and Young 2014), politics (Young and Keil 2014) and socio-economic structure (Hulchanski 2010).

In this chapter, then, I have built the case for a theoretical reorientation in urban studies. Propelled by both empirical developments that see continued settlement of the world's urban peripheries and by a critical reading of a Lefebvre-inspired urban theory, I argued for a de-centralization of urban thought and practice. The demand for an urban society, as hypothesized and predicted 'virtually' by Henri Lefebvre almost half a century ago, cannot be conceptualized anymore from the notion of the suburban deficit from which the 'right to the city' was developed. The centrality of the suburban form and society must be part of that reconsideration. The urban revolution, in this sense, does not have a central orientation but opens the city to (sub)urban society! In the following chapter, we will look at the ways in which suburbs have begun to be studied differently.

Zurich, Switzerland

4 | Suburban Studies

In contrast to the preceding chapter which discussed *theorizing* suburbanization, we are now more concerned with the mechanics of the study of suburbanization in academia and in practice. Suburban studies, as they are, have a reputation as a sub-region, a lesser domain of general urban studies. This can be laid at the feet less of those engaged in the study of urban peripheries than of those who have historically looked at the margins from the centre. In the long century during which we have now been involved in taking the studies of urbanization seriously, 'the suburban' as a real event has figured prominently while it has led a life in the shadows in the mainstream of urban analysis. Critical urban studies in particular have shown no love for suburbanization which has been looked at as a deviation of sorts, a strategy of capital shifting, a Haussmannian plan to expel the proletariat from the centre, or its inverse, the production of elite escape spaces, a process of cultural and ecological atrophication, etc. The sub-field itself has often reacted defensively and has delivered myriad historico-empirical studies but has remained out of the core theoretical debates that have driven urban studies overall.

In an insightful overview, Vaughan et al. (2009: 477) have identified four 'tacit assumptions that inform much of the historical and geographical literature [and] serve to obstruct a fuller conceptualization of suburban space': suburbs are non-problematic and one-dimensional; there is a teleology of dependent development that leads to the idea that suburbs have no development dynamics of their own; 'the belief that, as a powerful site of social reproduction, suburban space is adequately described normatively in terms of its multiple cultural constructions' between dystopian, realist and idealist ideas; and the suburb is a projection space of Otherness (from a form of earthly purgatory to locale of sub/urban revolt and insurgency). Vaughan and her co-authors conclude that 'There is, of course, no question that the suburbs do exist. Increasingly they are a theme of universal significance, implicated in the growth of globalized "world cities" and the rapid development of the built environment in emerging economies. However,…until the agency of suburban space as a distinctive domain of social organization is acknowledged, then the notion of the "suburb" remains too epistemologically fragile to carry the burden of representation that it currently bears' (2009: 485). This is a good place from which to look at suburban studies as a subfield of urban investigation.

A CALL FOR SUBURBAN STUDIES?

To avoid misunderstandings, it needs to be clearly said that the plea in this chapter for recognition of the planet's burgeoning peripheries as an object of study and source of theory that is not subordinate to the 'normal' centre, is not a call for a genre of 'suburban studies'. Instead, it is an argument for critical urban studies to be emphatically concerned with processes of suburbanization and suburbanisms as non-derivative processes in the move towards 'complete urbanization' (Lefebvre 2003). Such an extended urban studies takes the suburban as a historically evolving human geography in which more questions are posed than answers given. If being urban is increasingly the shared condition of our humanity, for many if not most of us, this takes place in what we would recognize as a suburban space. So why, and how, do we speak about suburbs and suburbanization today?

First and foremost we must avoid a replay of the Western colonial bias in urban studies. New developments and dynamics originating from the margins and peripheries defy the traditional dependencies of outsides from insides, suburbs from cities and expand our understanding of the dialectic of the urban process. Theorizing and researching, as well as acting upon suburbia, now needs re-situating in wider concerns for new geographies of urban theory and practice.

From this springs a wide-ranging field of research and practice, in these overlapping domains: Physical form/built environment; social relationships/governance; and (sub)urban political ecologies. These sections are roughly compatible with the classical sub/urban disciplinary approaches of planning, sociology/politics and geography/environmental studies. Three lenses lend further focus to these studies: governance, land and infrastructure: i. e. the dimensions through which suburbanization is driven forward, redefined and through which globally diverse suburbanisms are engendered.

Studies of the suburban have always been present in the pursuit of knowledge about cities and regions. Yet, they have often been peripheral to the core themes in the field. The gaze of urban researchers has been sometimes distracted to stray across the suburban expanse but it would invariably return to the centre of the city where insights on the essence of the urban were assumed to be had. This 'centre reflex', as we may call it, has eclipsed, for the most part, the formation of suburban constellations even in the second half of the twentieth century, when the production of space in the periphery was the driving force in the accumulation process in many capitalist societies. Nonetheless, a strong continuity of urban studies that focuses specifically on the suburban now exists. It has already contributed better to the understanding of metropolitan infrastructures, the production of land, regional governance and suburban ways of life. More recently, studies on the 'classical' suburban cases in Anglo-Saxon settler societies (USA, Canada, Australia) have been joined by broad examinations of the global variety of suburban constellations (Keil 2013; Herzog 2015; Harris and Vorms 2017).

Having thus moved the theoretical object of urban studies out from under the obscurity that covered its peripheries, we can return to the notion of an ongoing and one might argue accelerating, proliferation of urban society (Lefebvre 2003). From the early thinkers in urban theory, for example, Weber (1921), the Chicago School (Park, Burgess and McKenzie 1925), their Marxist challengers in the 1970s (Castells 1977), the city has been conceptualized from the centre outward. There have been counter-tendencies in Europe (e. g. Lefebvre 2003; Sieverts 2003) and in North America, especially around the Los Angeles School (Scott and Soja 1996; Soja 1996; for a recent overview see Judd and Simpson 2011), yet generally the peripheral, or suburban experience has been treated as dependent, or even deviant. Urbanization is still mostly imagined as a concentric expansion of centrality – of space, functions, economies, people. In turn, suburban studies (a large and growing field) have usually reproduced the centre-periphery split. Their disregard of the centre has been equally problematic (for recent overviews see Harris 2010; Forsyth 2012; Jauhiainen 2013; Clapson and Hutchison 2010). Suburban studies have largely shared the fate of their object as being seen as something of lesser importance. For every urban centre of relevance, there is a derived periphery. For each important work of urbanization, there exists a less weighty study of suburbanization. While commentary on urbanism as a way of life fills libraries and introductory lecture halls, suburbanisms as ways of life are barely recognized in their own right.

We might object, immediately, that this is surely an exaggeration. Suburban studies, as a field, have been around for a while, have made noted contributions and have produced masterful contributors. It has, many may argue, been as influential as the settlement form itself, suburbia, has been for the study of cities of the twentieth century in particular. Yes, certainly, the 'suburban solution' to capitalist overproduction crises, named so by a young Richard Walker (1977) in the 1970s, the suburbs have been at the core of not just how capitalist urbanization works but also how it fails. For many in the field, capitalism cannot be understood without its outlet in so much acreage of single home subdivisions; and urbanization cannot be understood without its (planned) Other at the periphery: sprawling suburbanization (Gottdiener 1977).

The suburbs have been around in our collective mind for a century at least. In North America, in particular, they have been the place of what Robert Fishman calls 'bourgeois utopias'. He cites Mumford's famous quip that the suburbs are 'a collective effort to live a private life' (1987: x). We are aware, of course, that this assumed right to a private life ended up being often more like a 'nightmare' than a utopian dream. The right was codified in the lasting institution of the 'restrictive covenant', a neighbourhood based contract of self-defined communities to restrict access of the Other from their home turf. Those covenants, Robert Fogelson writes, tell us about the suburbanites' 'fear of others, of racial minorities and poor people, once known as "the dangerous classes" and their fear of people like themselves. About their fear of change and their fear of the market, of which they were among the chief beneficiaries' (2005: 24). In Fishman's view, as he wrote about them in the 1980s, the suburbs were about to stay. While speaking about the 'fall of suburbia', he explicitly noted that he was 'not predicting a return to the cities' nor did he 'foresee grass growing in the gently-curving streets of abandoned subdivisions, nor the wind whistling through empty shopping malls' (1987: xi). Instead, the fall of the suburbs, in Fishman's view was equivalent to resulting in 'a new form of city' (1987: xi). This city, pace Fishman, was a collective elite creation and is, in his definition, a place of residential privilege.

The claim Fishman is making about 'a new form of city' is important as an attempt to remove the suburb from its marginality. Suburban 'accidental cities' of homes, malls and office parks, the 'boomburbs' that often get overlooked in the urban landscape of North America are in a process of mainstreaming, an awakening in the public mind. The 'suburban city', as Dolores Hayden (2003) called the phenomenon, is a form of normalization of what previously was outside of the core both geographically and cognitively. It also is an attempt to give space and credibility in research to aspects of suburbanization that had historically often been overlooked. While Fishman (1987), for example, is clear in his narrow definition of traditional suburbia as residential and single-family home based, others have noted that suburbanization was diverse from the start and included industrial, commercial and other forms of suburbanization that usually are underrated or ignored in histories of suburbia (Lewis 2004a; Lang and LeFurgy 2007: 10–11). Walker and Lewis consequently enlarge the definition of the suburban city to include more than residential land uses long before the birth of the technoburb: 'At the burgeoning edges of the metropolis we find a full panoply of workplaces, homes, infrastructure and commerce that make up the economy and life of the city' (2004: 31). Walker and Lewis also stand out from conventional suburban research in rolling their observations about the diversity of suburban histories into a more general quest for urban theory. Referencing in particular Mike Davis's (1995) take on urban theory used in the explanation of the growth of suburban Los Angeles under the spell of multiple disasters, Walker and Lewis argue: 'the combination of geographical industrialization, land development and metropolitan politics and planning is a theoretical framework that offers a means to advance beyond previous theories at the disposal of urban scholars' (2004: 31). While such broadening of the historical and empirical base for suburban studies is an important step and will yield important theoretical insights, it ultimately doesn't go far enough.

The suburbs are not just the future city but we can learn from the suburbs and their study something more general, perhaps even universal, about the urban overall without writing the suburban back into the political economy logics of urban economic development theories. As the owl of Minerva once again flies over suburbia, we can say in hindsight that Fishman was both right and wrong with his predictions about the fall of suburbia: the wind does now in fact blow through the malls of north America and grass grows in the yards of a post-subprime suburbia. And the fact that suburbia is now the city of record is not so much a consequence of its success as a particular form but because it is now more than a mere residential place of privilege: the suburbs are, in fact, a composite of all urban forms as they age into a post-suburban complex.

While Fishman stands out among suburban researchers in his broad and visionary approach to the subject, he is not alone. The most systematic examination of the suburban phenomenon has perhaps been performed by Canadian historical geographer Richard Harris. In his comprehensive life's work, he has subjected both the idiosyncratic self-built working-class suburbanism of the East End of Toronto and the mass-produced housing of the post Second World War years to detailed scrutiny. In his more recent work, he has expanded his perspective to the global scale as he has examined not just 'the material world' as Fishman would have it (1987: 9) of postcolonial South Africa and India but also the discursive reach of the concept of the suburban itself (Harris and Vorms 2017; see also Lang and LeFurgy 2007 on naming new suburban spaces). In a recent state-of-the art article, Harris (2010) provides typology of suburbanization. Remarkably, for a scholar who has made his mark predominantly through concise and thorough historical accounts of specific suburbanization processes and sites, Harris starts his taking of stock with a reference to Jenny Robinson's appeal that '[c]onsideration needs to be given to the difference [that] the diversity of cities makes to theory' (2010: 15). This appeal is both specifically postcolonial in its substantive significance and general in its call for a theory of urbanization that defies the normative pull of the dominant, the centre, the core, the citadel. In his own effort to challenge 'theory with diverse realities' Harris, quite in step with Robinson's original formulation, focuses on how 'the experience of the Global South [bears] upon our conceptualization of suburbs' and puts forth that there is a tendency, first identified by Robinson herself, to contextualize urban and suburban change there in a developmental frame. While there is much emerging literature in and on peripheral urbanization in the Global South, there is (a) little connection with the traditional suburban studies literature in the West and (b) surprisingly rare overlap among the burgeoning work on squatter settlements, gated communities, new middle-class suburbanization and so forth in the Global South (2010: 22–3). Invoking, then, a 'worldwide frame of reference', Harris (2010: 33) proposes a pre-theoretical typology of suburbs which differentiates in the first instance 'those who had some choice from those who did not and those moving towards as opposed to those leaving the city' (2010: 34). This distinction is pre-theoretical as it merely orders an empirical suburban world in meaningful ways; it is still taxonomy, not theory. These categories are immensely helpful to structure the kind of comparative work Robinson and others identify as non-negotiable parts of a theory building based on ordinary sub/urbanization. For Harris, this work flows into further, mostly empirical or taxonomic, research questions but not necessarily into theory per se. For that, suburban research needs to make connections to larger questions of social change, including the ones asked by critical urban theory that sees itself as social theory. This work has often been identified in recent decades with a Marxist, especially a Lefebvre-inflected approach to complete urbanization, or put differently, urban society. Kipfer, Saberi and Wieditz suggest that 'Lefebvre's most important contribution to social theory may lie in his ultimate decision (developed in the *Urban Revolution*) to place the urban in the middle of an open-ended social totality, as a level of reality in a mediating relationship to everyday life and state-bound and 'global' social institutions. Lefebvre's urban considerations play a constitutive, nonreductive role in the social order even as they refer back to lived experience and the state' (2013: 117).

SUBURBAN STUDIES: UNDER THE RADAR

This contribution was decidedly not based on a single-minded focus by Lefebvre on the unitary city in its European tradition but explicitly on his embrace of the suburbanization process, observed in Paris in the 1960s and 1970s as an inspiration for urban theory (Lefebvre and Ross 2015: 45–50). Some of the most insightful interventions in theory building from suburbanization comes from work done less on urban form, economic structure or political economy in the classical sense, from the study of suburban ways of life. In the research cluster on suburbanisms as part of our MCRI on Global Suburbanisms, two very different approaches to suburban ways of life have produced an impressive body of work with a large spectrum of methodological and conceptual approaches. At one end of the spectrum, we have systematic *quantitative* analysis of (Canadian) suburbanization using general categories of what might be seen as 'suburban'. These categories have been mapped out in ways that shatter and shift our understanding of suburbanization processes in many ways (Moos and Walter-Joseph 2017). Using 'single-family dwelling occupancy, homeownership and automobile commuting as indicators of suburban ways of living', Markus Moos and Pablo Mendez (2015: 1865), for example, have demonstrated that suburbs are becoming more diverse, 'consistently positive relationship between suburban ways of living and higher incomes' which they find troubling in a context of evolving bifurcation of housing policies in Canada. More importantly, perhaps, in the context of the current discussion, has been the theoretical provocation that was at the basis of this large-scale empirical exercise. This provocation was best summarized by Alan Walks (2013) who re-examined Lefebvre's theory of urban society. Lefebvre suggested in his 1970s projection of the future of urban life on the planet that the world witnessed an 'immense explosion' of urban form and life as the dialectical antithesis of the concentrated implosion and centralization experienced in the twentieth century. For Lefebvre, this dialectic of implosion–explosion and its accompanied complete urbanization of society was a mere projection in the 1970s, an idea whose time had come, a vision that only began to seriously materialize with yet another set of techno-economic revolutions at the turn of the twenty-first century. Be that as it may, suburbanization is a specific process of peripheralization, of explosion that can be isolated from 'the projection of numerous, disjunct fragments' among which Lefebvre also counts peripheries, vacation homes, satellite towns (2003: 14). But in contrast to the image of explosion which still implies the existence of a centre from where it originates, what we call suburbanization today is a broader dynamic that needs no dissolution of centrality into myriad peripheries anymore (as Lefebvre surmised in his multiple references to the subject). Suburbanization now proceeds through a generalized process of dissemination of suburban forms of life in a political economy of urban space which, as Fishman observed so brilliantly with reference to Fernand Braudel, is a 'transformer, intensifying the pace of change, [which] has moved from the urban core to the perimeter' (1987: 17). Walks focuses on the aspect of suburbanism in his examination of these processes and conceptualizes 'as an inherent aspect of urbanism that is both distinct yet inseparable from it – urbanism's internal ever-present anti-thesis that, in dialectical fashion, stands in productive tension with it, producing interleaved dimensions of "urbanism–suburbanism"' (2013: 1472). Importantly, Walks sees suburbanism therefore as 'a multidimensional evolving process within urbanism that is constantly fluctuating and pulsating as the flows producing its relational forms shift and overlap in space' (2013: 1472). Walks draws from this theoretical discussion six dimensions of urbanism–suburbanism which also lie at the basis of his and his colleagues' work on the quantitative analysis and representation of Canadian suburbanization (Moos and Walter-Joseph 2017).

At the other end of the methodological spectrum, Lisa Drummond and Danielle Labbé have proposed a theoretical approach to everyday suburbanism that is less based in quantitative data and shifting categories of general suburban and urban divides and more in an in-depth, anthropological and ethnographic view of suburban life at the global frontier of urbanization. Drummond and Labbé suggest that 'our studies of suburban everyday life...will require attention to the practices and the spaces of social interaction across the spectrum of suburban places and forms' found increasingly all around the world and especially outside the Anglo-Saxon suburban heartland '[a]cross the Global South, [where] cities are experiencing the construction, on outer urban edges, of informal peri-urban areas, new commercial developments, industrial compounds and middle-class housing estates which often sit cheek-by-jowl' (2013: 51).

In addition to and in concert with an expansion of critical suburban research into suburbanisms, we have seen an acknowledgement that suburbanization is now a global phenomenon (Keil 2013; Hamel and Keil 2015). This work builds on an explosion of regionally specific studies on peripheral urbanization in Eastern Europe (Hirt 2012; Stanilov and Sykora 2014a), Western Europe (Phelps 2017; Savini 2013; Tzaninis 2016), China (Fleischer 2010; Wu and Phelps 2008), Africa (Bloch 2015; Mabin 2013) and Latin America (Heinrichs and Nuissl 2015). In this work, we see a growing tendency of linking theoretical debates relevant to societal change more generally to the suburban phenomenon. For Stanilov and Sykora, for example, 'confronting suburbanization has become a mirror of confronting the wider societal challenges that need to be addressed in order to construct a sound framework for a sustainable future development' in post-socialist countries (2014b: 22).

SUBURBAN STUDIES REVISITED

Let us now return to suburban studies as originally evoked. Suburban studies is a proliferating field that carries with it a host of methodological approaches, disciplinary histories and definitions (Forsyth 2012; 2014). Much of it has been focused on morphology and the built environment and most of it, in fact by one measure sixty-seven per cent of recent literature, has been concerned with the United States alone (Harris 2010: 17) and as a consequence it, at times, has appeared as 'parochial'. Still, the field has global appeal (Jauhiainen 2013) and must be recognized as an important subfield in urban studies in its own right. As a field, it has often been defined through its empirical subject interests and less through its theoretical inquiry. Jauhiainen (2013) differentiates an interest in such themes as planned and unplanned, regulated and unregulated suburbanization, types of suburban form (terraced, villa, industrial and working class, garden, extended, gated, squatter settlements and the like). Those empirical study areas have resulted in an impressive body of work and some unfortunate self-referentiality (Harris 2010: 16). Jauhiainen identifies key quantitative, qualitative and symbolic changes to suburbanization in recent years that yearned for more and different modes of explanation. In calling for 'novel approaches and hybrid concepts' to explain the 'complex phenomenon' of suburbanization today, he notes these have to come from 'a range of economic, social and technological perspectives' (2013: 801). One should at least add a cultural and perhaps an environmental view to these perspectives.

Recent suburban studies have taken up this challenge in a variety of ways. A short review has to be prefaced with the observation that most of the new suburban studies continue down the road of the very American focus that has been identified by Harris above. In addition, most new suburban literature perpetuates the rather empiricist bent of most previous work in the field. The surge in suburban studies can at least partly be attributed to the institutionalization of suburban research in special centres or programmes and the funding of large research initiatives in the field. Perhaps most prominent among those efforts was the creation of the National Center for Suburban Studies at Hofstra University which has served as a noticeable and notable hub for researchers in the field. In one publication put out by the Center with the programmatic title *Redefining Suburban Studies*, Barbara Kelly writes: 'The field of suburban studies has for many years been anchored to that of urban studies, with the result that the suburbs have been studied as subsets of the city, rather than on their own merits. The idealized city is the standard against which suburbia is measured and found wanting. It is in all ways a lesser version, not quite as good, not quite as big, not quite as supportive as the city' (2009: 5).

Of the big narratives we can discern in the newer suburban literature one surely is entirely absurd and unproductive; the strenuous attempt by some urbanists to declare 'the end of suburbia' or name sprawl a 'dead end' (Gallagher 2013; Lutz 2017; Ross 2014). This tendentious and rather superficial work is based mostly on perceived trends of re-urbanization and sporadic suburban decline (malls and monster homes in particular) in some American cities. At its best, it is motivated by a strong belief in urban life, transit and public government. At its worst, it cheerleads gentrification of inner cities. It reifies a certain type of suburbia and declares its lessening importance equivalent to the end of a wave of peripheral expansion that is happening around the world. This position and the literature is part of the myopia that has characterized the field for some time.

Suburban culture has received more attention as diversity in suburbs increases. The international Cultures of the Suburbs project, for example, has examined 'cultural life of the suburbs' in a large number of suburban areas around the world (http://suburbs.exeter.ac.uk/). An exciting branch of the work on new diversities is actually also about how they intersect with traditional ecologies. Barraclough (2011) has, for example, examined the San Fernando Valley in Los Angeles as a product and project of a whiteness-inflected semi-rural culture with echoes of nationhood and exclusivity. Notions of particular appropriations of nature in the suburbs are central to the appellations inherent in this view. New cultural practices have also been subject to study among suburban researchers. A standout among recent studies is Rupa Huq's (2013; see also Webster 2000) vivid analysis of popular culture in and from (mostly British) suburbia. In the Canadian case, for example, Alison Bain (2013) has examined communities of cultural producers in suburban spaces. Cultural perspectives on the suburbs have always been central to the field. As research on the suburbs emerged in the post Second World

War years, this literature tended to be overdetermined by the kind of critical theory that was identified at the time with Horkheimer and Adorno's reading of American society lost in alienating massification and consumer culture (1972). The idea of blandness and mass cultural victimization of suburbanites became not just a scholarly truism, it populated general representation of suburbia from film to TV, music and literature. This attitude has finally shifted in recent years as the suburbs were becoming represented in more problematic fashion in films such as *Suburbia* (1984), *Get Out* (2017), or in TV series such as *Weeds* (Archer, Sandul and Solomonson 2015).

Suburban studies have recently experienced a shift towards an emphasis on rebuilding the suburbs by skipping the step of rethinking the suburbs. Only few examples exist of rethinking the suburbs theoretically, or at least conceptually in order to make them subject to planning or design action (prime and a bit of an exception among this practical work is the comprehensive 'manual' for suburban retrofit by Dunham-Jones and Williamson 2009; see also Jessen and Roost 2015a). Another systematic analysis in this vein is provided by Judith De Jong (2014) who uses the frame of 'flattening' to grasp the increasing breaking down of conceptual and morphological boundaries between suburbanizing inner cities and retrofitted suburbs. Yet, a big part of the work on suburbs is now taken up by urban planners and designers who have made the retrofitting of suburbs their primary concern without spending much time worrying about the theoretical implications of such rebuilding or their advocacy for it. As the decline of malls and other suburban forms and economic enterprises linked to a particular history of the society of organized consumption has taken hold, concern about the re-valorization of suburban land has engendered a cottage industry of design-related activities intent on urbanizing or re-urbanizing the suburbs. Some of this activity is additionally driven by considerations of reacting to challenges of climate change, sustainability and resilience for which more compact and centralized, less car-oriented forms of suburbanization have become *de rigueur* (Touati-Morel 2015). The retrofitting literature and practice fits neatly with the mainstream urbanist tendency of favouring certain city forms – mainly more compact, pedestrian oriented, centralizing in nature – over others – less dense, more auto or mass transit oriented and de-centralizing.

The dichotomies of city-suburb that underlie much of mainstream urbanist discourse and practice are insufficient. Suburbs are no simple and linear extensions of city cores but the product of a combination of dynamics – born from processes as diverse as anti-black racism, accumulation through the production of space, strategic investments towards just-in-time, flex-spatial, de-centralized regional economies and urbanist phantasies of renewal on the green field (just to name a few obvious ones). The post-suburban in-between city has developed its own logic and dialectics of space, contradictory and productive of new centre-periphery relationships beyond the old city-suburb binary (Quinby 2011). An important insight from this shift beyond the dichotomy is the notion that, as Bormann says, 'urbanity as a form of life has emancipated itself from the cities and has long nested in the urban hinterland, even in so-called rural areas'. We might then speak of a phenomenon of sub/urbanity instead. The process of generalization of sub/urbanity as a form of life works both ways: just as much as we now find stylized plazas and 'green rooms' in the lifestyle malls of exurban communities, the proximity to 'landscape', a traditionally non-urban trope, is now part of the definition of cityness itself (Bormann et al. 2015: 114–15; translation by RK). Retrofitting suburbia then operates on the assumption that bringing urban morphological diversity and urbanity into the suburbs will somehow bring improved everyday functionality, visual pleasure and vibrancy. But for retrofitting thus defined to be truly transformative and effective, it must first accept that suburbia itself is not monotonous but consists of various partial spaces that are ripe for change to reflect the demands of climate change adaptation, economic crisis management and societal shifts in demographic and family structure (Jessen and Roost

2015b). Density and compactness are seen as planning objectives that are achievable only in direct embrace with greenbelts and conservation policies (Keil and Macdonald 2016). This includes aspects of ecological and aesthetic reimagination of suburban and regional governance itself (Sieverts 2015).

Lastly, when we talk about retrofitting suburbia, we need to point to a distinct contradiction that appears most visible, perhaps, in the context of North American suburbanization, albeit more so in the United States than in Canada. Retrofitting the suburbs would need to be responsive to all of its constitutive landscapes of housing, commercial and office or manufacturing spaces (Jessen and Roost 2015b: 13). Yet, in reality, suburban landscapes, although they were the product of an unprecedented subsidy of general taxpayers to homeowners through housing loans and infrastructure construction, were ostensibly domains of privacy. This has had well-documented consequences for the sociality (or perhaps lack of it) of the suburbs and a certain leaning towards regional isolationism and a distinct drawbridge mentality. This point needs no further exposition here but it has consequences. The suburban single-family home and its privatist governmentality stands in tense contradiction to the modalities of government intervention and accumulation strategies (Ekers, Hamel and Keil 2012). Together these modalities have produced a landscape without apparent guidance in planning terms and without authority to do something about its obvious shortcomings (Bormann et al. 2015). This contradiction harks back to the very foundation of suburbia as a space in which original ideas of collective solutions were quickly subsumed under the poles of corporate profiteering on one end and a desire for privatism that was nurtured by consumer capitalism on the other. In a review of two new books on suburban history, Martin Filler (2016) explains that there is really no point in feeling nostalgic about or spending much time preserving the mid-century tract house, the very core of that privatized suburbia of which perhaps up to 35 million were built in the quarter century after 1945 in the United States alone. Viewing those houses as obsolete in terms of the family model they represent, their ecological unsustainability and structural flimsiness, Filler says, '[t]hey now seem more important in sociological rather than architectural terms' (2016). Historical alternatives for the production of suburban space, such as Robert E. Simon's plans for Reston, Virginia, were cast aside as the ideal of homeownership became more important than the necessity of providing housing in working communities. Instead of a suburban landscape of 'open spaces, homes and apartments that would be affordable to almost anyone, racially integrated, economically self-sustaining, pollution-free and rich in cultural and educational opportunities' (as Reston was described in Simon's obituary in the *New York Times*, quoted in Filler 2016), we now experience a tripartite confrontation of corporate commercialism, public–private 'exostructures' (Easterling 2014; Tonkiss 2013) and the hyper-networked private home. The incompatibility of the scales, logics and politics of these three interlocked landscapes of suburban building and retrofitting weighs heavily on the task of making suburbs more livable in today's building and rebuilding of suburbia continues to follow the path of infrastructure supply through government, short-term profit motives of corporations and developers and financial risk minimization of homeowners. The ingenuity and collectivist imagination of builders like Simon at Reston is mostly absent from these landscapes. Consequently, social, environmental, economic and spatial change suffers from multiple disjunctions.

QUO VADIS, SUBURBAN LITERATURE?

The literature in the field of new suburban research tends to be specializing rather than generalizing. No big claims are made. There is rather a tendency to advocate more local, specific and thematic approaches. Margaret Crawford (2015: 383–7) holds up D. J. Waldie's magisterial memoir *Holy Land* (1996) as a method of highlighting 'individual voice' and one example of what future suburban writing might aspire to.

Moreover, she identifies two more topics of research that she considers particularly 'promising:' habitation and suburban imaginaries. All three areas fulfill to a degree Robinson's and Harris's call for more ordinariness in urban research and theory building, but as it stands Crawford's call merely extends to more empirical research and not as much towards new theorizing on the city. In fact, although Crawford herself has brilliantly foregrounded the landscape of suburban Silicon Valley in her own research (2013), the reference point of such research remains the urban centre as its Other, whether the work is done in the area of biography, urban everyday life or imaginaries. Nonetheless, a shift is noticeable in the field, as yet of a pre-theoretical nature, but there is now a firm consensus developing that those studies that 'disregard suburbia's long and varied histories and the ways people have inhabited, revised and reinterpreted sub-urban areas over time' need to be rejected (Crawford 2015).

One exception to the trend in suburban studies to focus on more singular, limited events and places rather than larger general trends with theoretical implications has been the emerging literature on post-suburbia. Starting with the coining of the term in the 1990s (Kling, Olin and Poster 1995) and on the basis of work done by Fishman (1987) and Teaford (1997), there is now a general recognition that what's new about suburbia may be its maturation and complex patterns of change. A landscape ostensibly frozen in some ahistoric time, impenetrable to evolution was discovered to be just as susceptible to larger societal demands for adaptation as the city itself. Teaford describes 'the resulting post-suburban cityscape [as] simply suburbanization carried to the extreme, the end product of two centuries of continuous deconcentration of metropolitan population' (2011: 33). Where Teaford emphasizes 'continuity, rather than discontinuity' in the shift towards post-suburbia, others, like Tom Sieverts in his groundbreaking work on the *Zwischenstadt* (2003) have noted the new beginning that has characterized this period. Sieverts not only stresses the novelty of the in-between landscapes themselves but the necessity for experts and everyday suburbanites alike to cope with the newness of it all in conceptual and practical terms. Projected against a centuries-old history of the old city, the '[n]ew large-scale urban forms of a fragmented urbanized landscape [...] which didn't follow any planned urbanist ideas of order' posed novel challenges for planners, decision-makers and inhabitants (Sieverts 2011: 19). As these forms were once an image of a changing society, Sieverts notes the importance of a cultural and aesthetic approach rooted in the everyday practices and spatial appropriations of residents at a regional scale (2011; 2015). The scalar view of landscape (also recognized by Fishman 1987), is an important point in this literature on post-suburbia and the in-between city: While each subdivision, in fact each house in classical suburbia, could be studied in replicable isolation, the post-suburban landscapes of today don't allow such singular entry points. They always only make sense as an ensemble. In one extreme case, a group of architects and geographers has analysed an entire country, Switzerland, from the point of view of an urbanized, post-suburban landscape (Schmid 2014; Lehrer 1994; Phelps 2015).

Still, suburban studies have, for the most part, accepted their role as a subordinate, looking at a 'sub-set', as Kelly (2009: 9) puts it, of the urban problematique. In addition, there has been a tendency to reify the suburban in a variety of ways, as empirical object, as place or form, or as historical process, rarely as pathbreaking, multi-sited, socially complex phenomenon that might serve as a particularly privileged vantage point from which to observe and theorize urban society overall. Yet the limitations of the new suburban studies are just one element that impedes a more prominent role of suburbanization and the studies of the process in an overall process of theory-building. While this is not the place for a comprehensive review of such attempts at adjusting theoretical reach to a rapidly changing sub/urban world, a recent paper by Jamie Peck has identified a few of the important conundrums faced in theory building through comparison in a worldwide and world historic context.

He ends with an appeal for 'a continuing conversation across shifting terrains, dialogically conducted, not an act of spatialized inversion or a unidirectional "corrective"' (Peck 2015b: 179). Yet despite grounding his own work in the 'Vancouverist' landscape characterized by a densely centralized downtown jutting out from a suburbanized region (Peck, Siemiatycki and Wyly 2014), Peck fails to recognize explicitly the contribution the pervasive suburban landscape makes to our understanding of the urban and of urban theory (Peck 2015b).

Peck does not stand alone. After Walker's (1981) original intervention, critical urban studies, particularly in the political economy vein, has not paid much attention to suburbanization and, where it did, it has shown no love for suburbanization. The poststructural critique has not done much better on that account (see Harris's (2010) reading of Robinson's work for example). Ananya Roy doesn't explicitly mention suburbs or urban peripheries in her work on urbanisms. In her much-noted paper on new geographies of theory in *Regional Studies*, which Peck has in his cross-hairs with his follow-up piece in the same journal, she mentions suburbs and peripheries on several occasions but never specifically as a basis for theory-building as such. When she makes the case for a 'worlding' of urban studies, she talks about peripheries: 'A "worlding" of cities has now to take account of multiple cores and peripheries and more provocatively has to note the emergence of core–periphery structures within the global South'. She may have a different meaning of 'peripheries' in mind here in the first place but this statement can, indeed, be interpreted as an attempt to break open the static core-periphery (or more specifically also city-suburb) relationships in theoretically and conceptually meaningful ways. In fact, in a later contribution, Roy suggests just that: 'This is the task of de-centring (sub) urban theory, of not only fostering a sense of global urbanism but also of attending to the geopolitics of such globality. This too is at stake in the charting of global suburbanisms' (Roy 2015: 345).

It is possible, in closing, to distinguish two critiques of suburbanization. One is *systemic*, the other one *symptomatic*. The systemic critique sees suburbanization as an outgrowth of contradictions caused by capitalist urbanization processes and intrinsically entwined with them (Beauregard 2006; Harvey 2013; Smith 2002; Walker 1981). The symptomatic critiques largely disregard the systemic issues of suburbanization and see suburbs as a technical constellation that can be reformed and rebuilt into more economically appropriate, socially productive and environmentally sustainable forms. This second strand is largely driven by urbanist considerations (Gallagher 2013; Ross 2014; Sewell 2009). As we will remind ourselves throughout the remainder of this book, it makes a fundamental difference whether one studies suburbanization and suburbanisms as processes and relationalities of systemic significance (related to contradictions of capital accumulation, state formation, neoliberalization, racialization, etc.) or of mere symptomatic value (easily manipulated by choices in urban form, community economics, citizen behaviour, etc.).

Suburban Frontier, Palmdale, California, USA

5 From Lakewood to Ferguson

This chapter mainly does two things. It, firstly, unpacks one of the contradictions at the core of suburban studies, particularly in North America: that suburbs are the domain of a particular class of people, especially the white middle class. The contradiction lies in the fact that despite their alleged role as receptacle for what used to be called 'white flight', the suburbs (in North America) have been more diverse than commonly assumed. It may be, then, that the 'discovery' of the suburbs' diversity over the recent years is more a statement of changes in the politics and economies of suburban spaces and less about shifts in the statistically recognizable demographic make-up of the periphery. It also lumps spaces of privilege in with spaces of exclusion and confuses conversations about growing ethnic diversity with spaces of racialization. This first concern is motivated by debates about the suburbs as a place where communities of colour are increasingly concentrated. This is recognized as a shift from the 'chocolate city/vanilla suburbs', a paradigm, as one observer has put it, 'has almost entirely broken down' (Frey 2014).

The debate about the growing diversity of the periphery is bifurcated into one discourse that sees opportunity in emerging ethnoburban immigrant economies (Li 1998; 2009) even when that opportunity does not come without conflict (Pitter and Lorinc 2016) and another one that frames suburban communities of colour in a language of crisis, decay and danger. The voices that see the suburbanization of diversity as a positive outcome of urban change overall still often present it as a problem that is associated with the incoming non-white populations. An emerging consensus that 'White People Aren't Driving Growth in the Suburbs' over the next generation is often connected to an emphasis on 'new political and economic challenges as suburbs grow poorer and as whites adjust to (or rebel against) their new neighbors' (Capps 2015).

The chapter discusses, secondly, the ways in which suburbs have now become spaces of articulation of the global with specific places. This occurs in addition to the ongoing centralized spaces where globality is usually presumed to be located, but it also happens at the expense of those spaces. This debate both in the public domain and in academia intersects with the previous one considerably as processes of racialization and Othering are present in both. This chapter will seek to discuss the role peripheries play in defining globalization both through their importance to globalizing economies and due to their socio-demographic diversification. This second area of concern is motivated by what I consider an unproductive articulation of processes of globalization with spaces of centrality both in terms of the global North (as discussed by Robinson 2006 in particular) but also in terms of geographic metropolitan centres.

The chapter's title suggests that we will start our journey in a real and imagined space of simplicity where suburbs were homogeneous in terms of class, race and ethnicity. The Los Angeles suburb of Lakewood is taken as a symbolic marker in this debate. Lakewood was built as one of the first mass-produced suburbs of the post Second World War era, in the south bay region of Los Angeles. Marketed to (lower) middle-class white families, often veterans from the war, it was a place that symbolized segregated homogeneity and conformity (Waldie 1996).

The chapter ends with a discussion of African American suburbia using Ferguson, Missouri, as the conceptual and geographic anchor. That particular suburb of St. Louis where 18-year-old Michael Brown was killed by a white police officer in August of 2014 stands in contrast to Lakewood's sequestered whiteness as a place of exclusion, as a consequence of anti-Black racism and gentrification. Here is a description by Deborah Cowen and Nemoy Lewis:

> Ferguson has become a majority Black suburb because of gentrification and displacement in St. Louis and surrounding areas – though this shift is not reflected in the political leadership and the police force, which remain almost entirely white. The transformation of local social geography has not only been dramatic, but rapid. From a population that was 85% white in 1980, Ferguson had become 69% Black by 2010. (Cowen and Lewis 2016, n.p.)

Somewhere, between the guideposts of Lakewood and Ferguson, we can begin to understand the new diversity of the (American) suburb.

THE GLOBAL SUBURB: THE SUBURBAN GLOBAL CITY

Tim Bunnell and Anant Maringanti (2010) bemoan the tendency to see the most important financial metropoles, particularly those in the West, placed in the centre of research curiosity. This, they say, happens at the expense of other urban experiences that are left in the conceptual and empirical dark. The authors emphasize that it is necessary to see 'diversity as constitutive of contested global processes' and that we need 'to pay attention to the criss-crossing messy pathways through which ideas circulate, connecting cities in ways that can neither be ignored nor reduced to one-way traffic. The process of arriving at this idea conceptually and confronting the anxieties of dealing with the unpredictability of actual research requires building shared agendas and collaborative work practices among scholars at multiple locations' (2010: 418). Global cities in particular have mostly been imagined as spaces of centrality and as places in the global North and cultural West. Yet much of the globality of those spaces and places is located in or draws its strengths from metropolitan *peripheries*. In the debate on global urbanization, those spaces experience an additional intellectual and methodological peripheralization. They are considered secondary to the prime network spaces of the global economy. Yet the suburban has long played a role in the globalization of urbanity and deserves more attention. We need to address both the ubiquity of difference between central city and suburbs as well as the diversity within the latter, including their hybrid, in-between and post-suburban forms. Suburbanism includes elements such as centrality/peripherality, scale, mix of land uses, population characteristics, power/control, governance, mobility modes, services and amenities (Moos and Walter-Joseph 2017; Walks 2013). While one must consider their profound localized manifestation if transposing them across global suburbanisms, one needs to envision them as part of one world of the global city.

The global suburb – as a form of organizing sequestered peripheral space is not a recent phenomenon. It came with early modernity and the expansion of worldwide commercial, economic, religious and military interaction. The spread of capitalist metropolitanism from Europe across the Atlantic, into Africa and towards the East since the fifteenth century in particular entailed broadening the range of permanent settlements that were inhabited and used by multiple, previously unconnected communities from different corners of the world. Some of those new forms of settlement were 'suburban' in the sense of being spatially peripheral to existing urban cores and most displayed two aspects we nowadays associate with the globalization of city regions: the integration of centrally important economic functions and the demographic mixing of populations. One such historical example is the artificial island of Dejima in Nagasaki Bay which existed roughly between 1641 and 1853. The island, depicted in contemporary illustrations as crescent shaped and densely populated, served the rulers of Japan during their isolation from the external world in the Edo period as a location for import and export business, primarily with the Dutch but also with other European powers. Dejima which came to more widespread attention as the central location in David Mitchell's historical novel *The Thousand Autumns of Jacob de Zoet* (2010) can be read as a multifaceted metaphor: as a place of negotiation between cultures, as a hub of early economic globalization and as a relay station in the regulation and codification of international relations. But it can also, importantly in our context here, be read as an allegory for suburbanization: in classical manner, the island is a place for uses and people that have been pushed from the centre of the city into the periphery, in this case a maritime location. Dejima works as an early suburb of Nagasaki and as a basing point for the global economy. Subordinate to the centre of the Japanese city and an outpost in Dutch colonial trade, it is Othered, controlled and dependent. Still, the artificial island, equipped with the functionally diversified structures typical of an early modern settlement, is more cosmopolitan and more diverse than the central city and the metropolitan country that it serves. Dejima's diversity and function project today's reality of the global suburb which has become the centre of immigrant economies and cultures in many places around the world.

More than a century after the insular suburban existence of Dejima came to an end, as Japan and the Netherlands developed towards a different (post)colonial relationship, a polder, a stretch of land extracted from the sea northeast of Amsterdam replayed the suburbanization of diversity in much different circumstances. The *Flevopolder*, 'the largest artificial island in the world' (Tzaninis 2016: 20), has been home to the suburb of Almere for a generation. Designed in the tradition of the Garden City idea, Almere was thought of as an overflow vessel for Amsterdamers looking for living space outside of the congested central city. As Tzaninis notes, 'Almere's character has fluctuated over the past few decades between the typical suburbia of houses, picket fences and garages, and the new urbanity of experimental architecture, focus on consumerism and "smart growth"' (2016: 20). Most importantly, Almere has experienced 'an unprecedented demographic diversification for such a suburban new town'. Currently, almost a quarter of annual newcomers to Almere are immigrant (Tzaninis 2016: 70). As a consequence, or perhaps despite such diversity, twenty per cent of Almere's voters have recently supported the Islamophobic Party for Freedom (Tzaninis 2016: 21). *Like* Dejima, in previous centuries, the suburban island is a space of negotiation of (post)colonial relationships as the Netherlands, once a major player in the colonial geography of exploration and exploitation with ships on the coast of Japan, is now a magnet of diaspora communities that carry their aspirations across multiple immigrant pathways to the region of Amsterdam. *Unlike* Dejima, Almere is not a self-enclosed, tightly controlled space where populations from the entire world meet in uncomfortable proximity with little space to determine their own lives in a world ruled by providence and status – these points are illustrated vividly in Mitchell's main character, Jacob de Zoet, who struggles with the lack of ability to find his way in a world that shackles his romantic and career ambitions. Instead, Almere is a projection space for the fulfilling of individual immigrant life designs. It is where 'diversity exists "cheek by jowl": families live next door to singles, natives to immigrants, freelance professionals to manual labourers, asylum seekers to the privileged middle class, while flats are raised next to family houses and a Manhattan-esque centre looms next to the suburban homes' (Tzaninis 2016: 197). Yet both Dejima and Almere, both problematic spaces in a post-national world of upheaval, stand as variations on the theme of suburban diversity and globality, one that has been there from the beginning.

THE NORTH AMERICAN SUBURB INVENTED

With few exceptions (Gans 1967; Greason 2012; Kruse and Sugrue 2006; Johnson 2014; Nicolaides 2002; Richards 2012; Wiese 2004), the usual story we hear of suburban diversity is different, though. We have been told that suburbanization has in itself been a process of separating populations from a previous state of higher social mixity in inner cities. The term 'white flight' made an unequivocal statement on a racialized process of segregation whereby white (and the implication was 'middle class') residents intentionally moved away from black (working-class) residents to take up residence in the suburbs. The homogeneity of form often enhanced the impression of homogeneity of population. D. J. Waldie, resident of one of the most iconic of those post World War suburbs of homogeneity[1], Lakewood south of Los Angeles, California, describes the aerial photographs taken of his emergent hometown during the construction phase in the 1950s: 'The black-and-white photographs show immense abstractions on ground the color of the full moon.... The photographs celebrate house frames precise as cells in a hive and stucco walls fragile as an unearthed bone. Seen from above, the grid is beautiful and terrible' (Waldie 1996: 6). Waldie notes that the 'terrible' homogeneity of Lakewood's suburban grid (which he points out was working class, rather than middle class) led observers at the time to speculate that 'the grid, briefly empty of associations, is just a pattern predicting itself' (Waldie 1996: 6) and he bemoans that the 'theorists and critics did not look again, forty years later, to see the intersections or calculate in them the joining of interests, limited but attainable, like the leasing of chain stores in a shopping mall' (Waldie 1996: 6).

Shortly after its inception as the second such suburb after the East Coast's Levittown in 1954, Lakewood became a new model of suburban political self-governance that did indeed create the condition for subsequent segregation by class and race that proved to be the foundation of many of Southern California's persisting inequalities. White, mostly well-to-do, communities 'privatized with class' as Hoch so aptly called it (Hoch 1985). This pattern that sharply segregated communities by class and race and laid the groundwork, among other things, for the highly detrimental tax initiative Proposition 13 in the 1970s, was indeed corrosive and ultimately led to much redistribution of social wealth from working-class communities of colour to middle-class white communities.

FORESHADOWING THE GLOBAL SUBURB IN THE SUBURBANIZED CITY

Yet, Waldie's portrayal of Lakewood also fits into another narrative pointing towards the originality and continuity of suburbanization in Los Angeles. The Lakewood developers 'planned to build something that was not exactly a city' (Waldie 1996: 4). Los Angeles has always been suburban, though not in the sense that an existing centre had suburbs: ever since the 1880s, the city exploded in a way that scholars and other observers would eventually name suburban (Warner 1972). In the 1920s, the city had already defied all boundaries of urban meaning. But that it has always been suburban is really captured in perhaps the most crucial period of its existence when it became known less as an aberration and more as a model of future urban form and social structure. The step from talking about suburbanity as a mere phenomenon of note to recognizing its central significance for the way this city and perhaps most cities are now put together was not taken only by the authors of the Los Angeles School and their epigones over the past two decades. In that sense, Lakewood was not just a symbolic space of separation and sameness but also the possibility of difference spatialized in new ways and perhaps a premonition of the kind of mosaic of post-suburban diversity we experience today. Tzaninis calls this 'kaleidoscopic suburbia' (2016: 20).

This premonition of ubiquitous diverse utopia in Los Angeles revealed itself first most prominently in the 1960s, when the smog was unbearable, when the United States and the world learned about a one-time riotous suburb called Watts and when Charles Manson's gang murdered Sharon Tate and others in the Hollywood Hills and Los Feliz. This was also the time when Antonioni made *Zabriskie Point* (1970) which introduced the limitless urban region through the depiction of 'a vicious circle in which the escape from Los Angeles will always lead us back to the city' (Keil 1998: 40). The best known outsider who noticed that Los Angeles at the time was becoming more than an oddity but rather a leader in urban form and function, was Reyner Banham who himself was influenced strongly by another itinerant observer almost half a century before him, the German geographer Anton Wagner, whose meticulous study *Los Angeles, A City of Two Million in Southern California* (1935), laid the groundwork for subsequent revisionist ideas on the type of urbanization that might be in store for the twentieth century. In its own right, Banham's *Los Angeles: The Architecture of Four Ecologies* (1970; see also Fontenot 2015; Marshall 2016) revolutionized the way the suburban landscape could be viewed: disdain for and judgement about it was replaced by curiosity and explanation. Driving on a freeway became a more normalized urban activity and living in a suburb was part of the overall metropolitan experience rather than a social pathology.

What seemed to shock some conservatively Chicago School-trained urban studies scholars in the 1980s, when the Los Angeles School began its aggressive pushback of the centralist tradition in the field was, in fact, not at all new. Even casual observers noted as far back as the 1960s that the city had evolved into an ongoing real life experiment on how living together in urban space could, and often would, be advanced in fully automobilized societies. A French television station aired a brief film called *Anatomy of Los Angeles* in 1969. Its French voiceover makes some important observations on the way that Los Angeles foreshadowed the urban future:

> At the feet of this kingdom another decorative city bustles about and whispers. A blue, flat city: Los Angeles. Fractured into multiple working-class areas that ignore each other, inhabited by individuals who live together but never meet, a city wedged between the desert and the ocean, constantly under threat, its heaving heart torn, dislodged – deprived of

a center by the existence of the desert. Los Angeles has a surface area half that of Belgium. The people of Los Angeles love to think that they live in the city of the future, but it is rather a city of the present incarnation of past times and as a consequence the very thing that unknowingly empties our cities once the outskirts begin to grow. If you would like to know what the outskirts of Paris, London, or even Moscow will soon look like – what their problems will be, what is waiting for us, threatening us – you must go to Los Angeles. It is a true urban laboratory. Entire districts are built suddenly; others are knocked down and rebuilt. The primitive desert is like a black canvas on which suggestions for settlements are drawn and erased straight after, as if with an enormous cloth. But it is not just about walls, it is about people who try out their existence and then pass on to another one, like actors who pass from one role to the next. A population laboratory. [2]

Remarkable are two things: First, the astonishing precision with which the short passage names both the suburban character of the city and the effect it has on the dialectics of centrality and de-centralization; and second, the passage calmly and self-confidently predicts that the placeless non-centrality of Los Angeles will soon be shared by the 'outskirts of Paris, London, or even Moscow'. The audacity of this claim must have been outrageous at the time as all these cities were *inward* looking despite the fact that they all expanded aggressively *outward* themselves. But the idea that the standard of urbanization worldwide might at one point be set by the sprawling expanse of a desert frontier town was unimaginable to most. Henri Lefebvre, foremost among the urban theory giants of the 1960s and 1970s noted after visiting the city in the 1980s: 'There is something stupendous and fascinating about [Los Angeles]. You are and yet are not in the city. You cross a series of mountains and you are still in the city, but you don't know when you are entering it or leaving it' (Lefebvre 1996; cited in Keil 1998: 50).

Positioning Dejima and Almere in the history of Dutch colonialism and Lakewood in the geography of Los Angeles's sub/urbanism allows us to raise the dialectics of continuity and change that situates sameness and difference in suburbia. We all have grown up to various declinations of the suburban dream, whether we lived it ourselves as our families settled on the periphery or whether we lived somewhere else but were treated to various versions of televised suburbs in the soap operas of our childhood and youth. D. J. Waldie addresses his reader directly: 'You and I grew up in these neighborhoods when they were an interleaving of houses and fields that were soon to be filled with more houses' (1996: 3).

The Lakewood myth of sameness and uniformity also rests on a particular myopia that we discuss in more depth in the chapter on suburban infrastructure that will follow. Suburbs are treated largely as singular and unconnected. Their 'purity' and 'insularity' is more an ideological figure than a representation of their reality. In fact, their functioning relies on a complex commercial and industrial landscape that belies the image of the allegedly homogeneous residential suburb. The entwined metabolisms of suburban life are hidden behind the veneer of separation. Still, as we will see, things have changed.

SUBURBIA RECAST

Of late, we have witnessed that, ever so slowly, all the good things we thought were to be found in the suburbs – walkable cul-de-sacs filled with cycling kids, green nature, free mobility, etc. – have been turned into their opposite. Once they were considered landscapes of middle-class desire, these monotonous, monocultural domains of automobility, had soon caught the attention of the 'urbanist' planners and academics who deemed them uninteresting at best, pathological at worst. But they also become the object of scorn and fodder for the ennui of suburban youth and homegrown artists alike. The Canadian rock band *Rush* put this into words at the height of the suburban building boom in 1982 by calling out the suburbs' 'geometric order' and 'pre-decided future'. Their 'insulated border' reflected the suburbs' own boundaries, external and internal and that was supposed to be part of their appeal. But it all couldn't be contained. Change came to the periphery.

THE SUBURBS INVERTED

Alan Ehrenhalt (2012) argues that inner cities in the United States are experiencing a partial revival after decades of decline, rising crime and poverty rates, etc. and the suburbs, both inner and outer, are the site of new kinds of poverty and diversity. The admission or discovery of growing poverty in American suburbia is a mixed narrative where statistical information vies with various reconceptualizations of the suburban-urban divide (Pooley 2015; Semuels 2015; Vicino 2008). Concentrations of poverty are considered a particularly pernicious problem in cities and suburbs (Kneebone and Nadeau 2015: 15). And poverty as well as extreme-poverty has grown extraordinarily fast in suburban census areas in the United States (Kneebone and Nadeau 2015: 29) where 'by 2005–2009 the number of residents living in extreme-poverty neighborhoods had risen by a third compared to 1990' (Kneebone and Nadeau 2015: 35).[3] Yet, poverty of households and people is often confused with spatial concepts like 'poverty by postal code' or morphological 'vertical poverty' referring to (suburban) high-rise neighbourhoods[4]. The classification of some areas in urban regions as poor has its roots in the earliest efforts of urban sociologists in the human ecology tradition and in the statistical rationality of the state. It was, indeed, unusual until recently to look for 'deprived' or 'declining' neighbourhoods in the outer reaches of metropolitan areas, even in Canada, where much suburbanization, especially in high-rise tower neighbourhoods, immediately raised concerns about social problems. In addition, as has been observed in many locales from Eastern Europe to France and Canada, the morphology of suburban tower neighbourhoods has played a key role in the narrative of the failure of the welfare state and modernist city planning and sometimes all attempts by government to regulate or initiate housing for the poor. In Toronto, the suburban towers 'stand as large-scale representations of the devalorization of the inner suburbs: drab, out of fashion and in need of renewal' (Poppe and Young 2015: 616). Accordingly, in Toronto, the inner suburbs have been singled out as the locale of persistent, chronic and worsening household poverty and have been the target of place-based policies to right the wrongs of large-scale socio-spatial polarization (Hulchanski 2010; see also Ley and Lynch 2012).

American suburbs in the early 2000s had roughly the same demographic mix as central cities did in 1980 (*The Economist* 2008). Suburbs 'are no longer places with high proportions of home-owning, non-Hispanic Whites and native borns with relatively high household incomes, high levels of education and without any problems' (Anacker 2015: 1). A growing number of case studies in individual cities in the United States and Canada and increasingly also in Europe are pointing to the phenomenon of increasing suburban diversity. Far-reaching, national quantitative studies (such as those by Kneebone and Berube 2014 and Kneebone and Nadeau 2015) have led to a better understanding of the fine grained social and economic differentiation now found in suburbs and have inspired classification of vulnerability to social decline such as 'rapid growth', 'strained', 'at risk' and 'distressed' suburbs in various jurisdictions across the United States (Kneebone and Berube 2014). Some observers have warned, though, that we should not expect too much mix and diversity in the new diverse suburbs: the segregationist patterns that plague American society elsewhere are also noticeable in the periphery. John Logan, who has systematically studied long-range demographic change in American metropolitan areas, noted accordingly: 'Suburban diversity does not mean that neighborhoods within suburbia are diverse' (Logan 2014) and 'old urban forms of inequality are replicating themselves in the suburbs' (Badger 2014). This affects, for example, the black middle class that, when arriving in the suburbs, experience what Elijah Anderson has called 're-segregation' (2015).

The debate about the new suburban poverty needs to be contextualized in a discursive shift in the way cities have come to be seen. The suburbs, once considered a solution to many of the city's social, hygienic, environmental and cultural issues, have been increasingly cast as problematic for reasons of their morphology or design and sometimes because of the people who live there. A typical notion that is pervasive in this context is that suburbia 'wasn't built for poor people': 'Designed around a car-centric culture of single-family homes clustered in cul-de-sacs served by strip centers and shopping malls and fuelled by jobs reached by commuting to downtown or suburban office parks, suburbs…have struggled to respond to denser populations, increased congestion and, as a result of the 2008 recession, a decline in the middle-class jobs that made it all possible' (Burns 2014). The idea of poverty moving to the suburbs reached a climax when, even before the full onset of the recession of 2008, in light of large-scale loss of value and the abandonment of many houses in new-built exurban subdivisions, the viability of suburban life was increasingly being questioned. An article by Christopher Leinberger from the March 2008 issue of the *Atlantic* sounded the alarm when it speculated that while acknowledging the particular problems in inner suburbs, 'much of the future decline is likely to occur on the fringes, far away from the central city, not served by rail transit and lacking any real core' (2008: 73). Images of rental apartments in subdivided former single-family houses and even houses squatted by drug users and homeless people on the exurban periphery were not unusual in this discourse of moral panic at the heart of the American Dream (Leinberger 2008). As Alex Schafran has pointed out, the very discourse of suburban poverty has been problematic as 'commentators have reached for the pejorative urban language upon which generations of Americans have been weaned – slum, blight, ghetto – and begun unleashing them on the suburbs' (2013: 130–1). Critiquing the language of 'slumburbia', Schafran (2013) also notes that this 'analysis' keeps operating in a traditional dichotomy of city and suburb just when most scholars were beginning to see continuities of a post-suburban landscape evolve: 'scholars of suburban poverty and the deeper structural problems of the metropolitan fringe should be wary of the concept of 'suburban decline'. It is all too easy to fall into a means of defining decline based on who lives there, or even worse on who has moved there' (Schafran 2013: 143; see also Hesse 2015).

MAKING SUBURBS DIVERSE

While the debate on the 'great inversion' (Ehrenhalt 2012) is not necessarily an entirely new phenomenon (Bloch 1994; Harris 2013) and is certainly not as important in European contexts, in Canada or in the Global South where peripheries have been more mixed all along, it has led to an increased interest in questions of new suburban diversities, be they ethnocultural or socio-economic. A growing literature has begun to explore those diversities with methodologies that span from the quantitative (Anacker 2015; Hanlon, Short and Vicino 2010) to the ethnographic (Cheng 2014). The turn towards recognizing the suburbs as culturally diverse has been around in the social sciences at least since Li's (1998; 2009) work on *ethnoburbs* that explored the diverse economies and social structures of immigrant communities in cities like Los Angeles and Toronto. In Toronto, for example, we observed elsewhere that 'a new type of enclave is developing in these ethnoburbs, one in which new communal structures are evident, often in the context of religious institutions but also related to specific industrial and labour market sectors' (Allahwala and Keil 2012: 125). We are now entering a phase of majority minority suburbia. Margaret Crawford, in a recent summary of suburban trends, goes so far as to postulate 'that difference may actually be the defining characteristic of suburbia, rather than the sameness consistently attributed to it. In fact, currently, in an inversion of conventional wisdom, cities are becoming more homogeneous while suburbs grow more diverse' (Crawford 2015: 382).

The inside view of suburbia is as central to the maintenance of suburbanism as a way of life as is securing the perimeter – with lot sizes, tax regimes or gated walls. On the inside, conflict needs regulation in proximity, not at a distance: 'In the suburbs, a manageable life depends on a compact among neighbors. The unspoken agreement is an honest hypocrisy' (Waldie 1996: 20). Wendy Cheng, who has examined Southern California suburbia of a different generation from Waldie's Lakewood, notes that the life histories of that area's residents 'in the contemporary moment challenge the old trinity of whiteness, suburban and American – both what we may think that each of these terms means, as well as their easy conflation with one another' (Cheng 2014: 21–2). Cheng concludes on the basis of her research in the San Gabriel Valley east of Los Angeles that 'culturally, ideologically and politically, the voices [in her book] unsettle the United States' long-held image of itself as a white suburban nation, with significant implications for the presumed conditions and terms through which people form class, racial and national identities' (Cheng 2014: 22). Cheng sees a 'moral geography of differentiated space' at work among Asian American and Latin communities that have been populating the eastern suburbs of the Los Angeles region (Cheng 2014: 197), which points to a different human ecology than the one suggested by the original Lakewood model of separation. Sandwiched between white privilege and anti-Black racism on the housing market, 'Asian Americans and Mexican Americans took advantage of the availability of diverse housing options' in the San Gabriel Valley's suburban expanse. Before the emergence of 'ethnoburbs' of new Chinese immigrants, there was multicultural arrangement in that part of the region. Always challenged by competing racializing discourses that define and delimit communities in the United States, those suburbs had long become a place of intense redefinition of the assumed colour lines that define exclusion in that region or nation.

THE TORONTO COSMOPOLIS: FROM SCARBERIA TO BROWNTOWN

It is not accidental that the Toronto pop group Barenaked Ladies set their haunting cover of Bruce Cockburn's ballad 'Lovers in Dangerous Times', in a bleak wintery Scarborough, an eastern suburb of Toronto which was built in post Second World War years and originally predominantly populated by the working-class families who found jobs at the Fordist factories and assembly lines of an ascendant industrial economy in complexes such as the Golden Mile. Housed in bungalows and high-rise apartments, these mostly European immigrants came close to living the ideal of a Keynesian growth economy with safe and stable employment, unionization and for many also, homeownership. With the end of the Fordist mass production period and the closing of the plants, the area experienced a socio-economic and socio-demographic shift. As the good manufacturing jobs disappeared, largely replaced by lower wage, less secure, usually part-time service employment in retail, restaurants and other sectors, immigrants from non-European countries who have been the majority of all newcomers since the turn of the century (*Economist* 2016) moved massively into the hundreds of rental apartment towers and relatively affordable bungalows and have provided the service labour power of globalized Toronto. The draw of Toronto's suburbs has been an important backdrop of the story of arrival cities (Saunders 2011) and the subject of intense interest for urban scholars in the region (Ahmed-Ullah 2016; Poppe and Young 2015). This work both builds on and moves beyond the stereotypical application of viewing immigrant suburbs as a copy of the French *banlieue* and a version of a displaced ghetto (Dikeç 2007; Enright 2016; Wacquant 2008). Perhaps most importantly, these developments trouble the conventional use of the concepts of multiculturalism and diversity that seem innate to Canadian urbanism. As Jay Pitter tells us in her introduction to a critical new book on the subject, we have entered a period of 'hyper-diversity'. Politically, this entails an 'intersectionality [that] requires us to acknowledge that our cities contain diversities within diversities within diversities. They are deeply complicated places' (Pitter 2016: 9). One of those newly complicated suburban places 'where the visible minorities are now the majority' (Ahmed-Ullah 2016: 242) is the City of Brampton, home to almost 600,000, in the north west of the urban area. More than 350,000 of these were citizens from visible minority populations in 2011. Noreen Ahmed-Ullah lists nicknames for the place as 'Browntown', 'Bramladesh', or 'Singhdale'. These are reflections of a majority South Asian residential population and perhaps also workforce, in the burgeoning suburb whose population is projected to grow to 840,000 in 2031. Brampton has now entered debates about its future that entail, among other aspects, the question whether the presence of forty per cent South Asians in one place will eventually lead to ghettoization or enclaving in the future. Be that as it may and only speculative answers can be given at this point, a new era has begun. Ahmed-Ullah resumes: 'Regardless of whether Brampton is or isn't a ghetto, that label alone is an ominous marker. It raises tough questions about the future of a city that's been profoundly reshaped by the immigrants who've made their homes here' (Ahmed-Ullah 2016: 244). To Ahmed-Ullah there could be all kinds of answers to these tough questions, from whether this leads to more insularity or more political power, for example. One thing is clear, though, if we want to understand Canadian suburbia, or for that matter the Canadian city today, we had better start in the urban periphery where we find Brampton, Mississauga, or Markham, all suburban municipalities with majority 'visible minority' populations. The hopeful depiction of suburbs of aspiration, embedded in an alternative narrative of Canadian immigration and suburbanization, also stands in contrast to a story of decline, decay, exception and violence that has been taking shape in the diverse working-class suburbs of the United States. The introduction to that narrative, though, was written by another Canadian rock band.

THE ROAD TO FERGUSON

In 2010, Montreal indie rockers Arcade Fire, released a theme album called The Suburbs, which was characterized by a decidedly darker vision of the urban periphery. In their title song, they refer to future 'fighting/In a suburban war'. Spike Jonze's video for this concept album displayed a thoroughly dystopian backdrop to the rather upbeat music by the Arcade Fire. Jonze conjures up images of civil war and authoritarian state action: Tanks are rolling along leafy streets. The suburbs in or *as* a state of emergency was not common fare outside of radical critiques of suburban landscapes such as Mike Davis's grim depiction of Los Angeles in the late twentieth century, especially the racially motivated police brutality against youth of colour in that city's diverse suburbs in the 1980s and 1990s (1990). Turning bourgeois utopia into the dystopia of an occupied territory, in Jonze's video experienced by white middle-class youth as a state of exception, has been a perpetual reality for kids growing up non-white in the inner suburbs between Los Angeles and Toronto, Chicago and New York, or Paris, Frankfurt, Lisbon or London for that matter.

Americans and the world recently were able to see this fictional account of a suburb 'gone bad' foreshadow the real developments around the shooting death of African American youth Michael Brown at the hand of a Ferguson police officer which set off days of protests and sent shock waves around the nation and the world. There we witnessed a militarized police force fighting its own citizens with war-tested equipment, in the streets of a damaged but still recognizable post Second World War suburbia. Ferguson, a suburb of St. Louis, Missouri has since been subject to intense scrutiny and has, by some accounts, become the showcase of a new type of suburb which is more diverse, poorer and a lot less defined by self-controlled boundaries or life-style preferences which was always part of the common image of what suburbs were defined as. The story of Ferguson as a *place* that entered the collective consciousness in the summer of 2014 reveals a significant blind spot among the American and general public when it comes to the geography of race in the United States. French, English or German observers, for example, that have seen the suburbanization of racialized populations for generations, would not have been inattentive to the discoveries that were made in the wake of Michael Brown's murder.[5]

Just as they did in 1965, when the Los Angeles suburb of Watts exploded onto the national scene and white Angelenos allegedly went on the Harbour freeway to view their invisible ghetto that they had not taken notice of before, the mainstream (white) media and public expressed astonishment that poor African Americans could, indeed, be found in the metropolitan fringe. Even liberal (and critical) commentators like Peter Dreier and Todd Swanstrom are exasperated with surprise to find misery in the periphery: 'Ferguson is a suburb. More specifically, it's a suburban ghetto. Today, about forty per cent of the nation's 46 million poor live in suburbs, up from twenty per cent in 1970. These communities (often inner-ring suburbs) are beset with problems once associated with big cities: unemployment (especially among young men), crime, homelessness and inadequate schools and public services. Their populations are disproportionately black and Latino' (Dreier and Swanstrom 2014). The immediate and seamless discursive shift of the terminology of the 'ghetto' from its traditional inner-city locale to the urban margin is as troubling as the premonition that Ferguson is a 'microcosm' of larger 'problems' that have beset the American metropolis. Isolating issues of political representation and systematic impoverishment, Dreier and Swanstrom make a host of recommendations for policy and action that may help communities like Ferguson to get on their feet again (Dreier and Swanstrom 2014).

Of course, the narrative of surprise when it comes to the racialized ghetto is a hard sell among African Americans who have lived suburban lives for decades. It is also not at all news to those who have seen African Americans leave central cities under pressure of gentrification and white population growth in neighbourhoods close to the city centre. Formerly black majority cities in the United States have been bleeding black residents in many metros in the US (Saunders 2015). But once again, it is not the purpose of this chapter to engage the statistical minutiae of population movements across and beyond metros in the United States. We are concerned here, instead, with the discourses that attach themselves to the events associated with places 'like Ferguson'.

One such place 'like Ferguson' is Compton, a Los Angeles suburb just north of Lakewood. Here a different discourse of suburban identity formation has been taking hold for a generation. Compton, one of the poorest suburbs in the United States became, for a while, the unrivalled world headquarters of hiphop music. When the rap group N. W. A. sold half a million copies of its album 'Straight Outta Compton' in the summer of 1988, a flood of defiant self-stylization was triggered in the world-forsaken community. The creation of communal identity was achieved through a decidedly anti-establishmentarian attitude that expressed itself in blunt realist style and with an unapologetically capitalist sales strategy fuelling an entire industry of Compton marketing. The rappers of Compton subscribed to an ostensibly unmediated representation of reality and truth expressing the identity of place in Compton. The reality strategy aimed at the production of visibility in a situation of complete marginalization and criminalization. While other suburban local states marketed themselves as localities of innovation, stability and success, Compton's global music market handlers sold the city globally as the home of the drive-by-shooting: 'It's the Compton Thang.'

A generation after N. W. A., a new artist from Compton, Kendrick Lamar, reflects on this legacy in a song, co-performed with N. W. A.'s

Dr. Dre with the title 'Compton' (2013). Lamar laments that while Compton remains a city unlike any other, the world of international hiphop owes the place much of its overwhelming success. In the new Compton of Kendrick Lamar, the suburb has become just as much a fictional reality that contains all of the (now sampled) dreams and nightmares of the artform of hiphop and a new municipal reality in which a majority African American suburb takes control of 'heritage', the wind in the back of consecutive Obama administrations and a set of new connectivities between the critical suburban rail infrastructure in the Alameda corridor and the Blue Line rapid transit that transect the South Bay. Self-determination here goes beyond classical notions of representation although systematic impoverishment does persist.[6]

Returning, then to Ferguson, from a more critical perspective, we can develop a different and more acute narrative of African American suburbanization. Positioning themselves in a more radical vantage point, Cowen and Lewis (2016) note: 'Black suburbanization emerges largely in response to gentrification of inner-city neighborhoods and the displacement which ensues. It is a striking feature of the geopolitical economies at work in a place like Ferguson, Missouri.' Cowen and Lewis don't just point to gentrification at the basis of the malaise of African American Ferguson, but they also point to a long history of a geopolitics of empire at the root of anti-Black racism that underlies it (see also Keil 2007). And in contrast to Dreier and Swanstrom, these authors *problematize* the category of the 'ghetto' rather than deploying the controversial term superficially to Black suburbia.

Lastly and returning to our opening representation of Lakewood as a safe space, a white working-class utopia sequestered from not just the real inner city but from the idea of the inner city itself in the 1950s, there is nothing safe about black suburban Ferguson: 'The policing of Black suburbs in Missouri shatters the myth of suburban safety, suggesting that police violence can be even more severe because of the threat Black suburbanization is felt to pose to suburban spaces – so deeply associated with whiteness. What is distinct about this moment is clearly not that police violently oppress Black people who are already imagined as either or both a threat to the social order or a resource for it. Rather a more precise but also much more diffuse set of transformations in the domestic violence of the state operates today through the financialization of municipal government and urban space' (Cowen and Lewis 2016).

Black suburbanization, then, is not entirely new and it is not the geographical transplantation of 'the ghetto' although the generally assumed pattern that white immigrants were able to leave the slum while African Americans were stuck in racially defined areas (Philpott 1978) also pertained to the formation of suburban spaces such as those in formerly agricultural New Jersey (Greason 2012). In fact, African American suburbanization followed in the same pattern of development that had confined them to marginal settlement spaces all along: in dangerous, vulnerable lowlands, areas between major infrastructures and close to toxic or noxious facilities: 'on the dangerous land that Whites would not – indeed could not – inhabit, the Black settlers found a way to make the inhospitable hospitable' (Newkirk 2015). Pulido has noted that 'suburbanization can be seen as a form of white privilege, as it allowed whites to live in inexpensive, clean, residential environments' (2000: 16). These processes involve what Elaine Lewinnek (2014) in her book on Chicago's suburbanization calls 'mortgage of whiteness:' the defense of white working-class communities of their houses as economic assets, part of their 'reward' in a booming capitalist shock city that redistributes economic wealth through racialized and classed space. The process was, as Bill Savage points out in a review of Lewinnek's book, as violent as it was systemic: 'As racist real-estate covenants and mob violence kept African-Americans trapped in the Black Belt, overcrowding produced blight and slum conditions that whites then used to justify not allowing blacks to buy into their own neighborhoods' (Savage 2014). In other instances, like in Los Angeles, the white working class initially benefitted from employment in and residence close to industrial suburbs. In later decades, as these suburbs 'matured', people of colour began to inhabit those areas that were often marred with compounded environmental problems linked to their industrial past. Hence one might add that Black suburbanization is also a product of longstanding processes of spatializing white privilege as it became part of 'ethnic succession' when white working-class families moved from inner-ring suburbs further out (Pulido 2000; 2015). This spatialization has had long lasting consequences; it influenced the politics of, and after, the civil rights movement because white identity and suburbanization were merged into one after the Second World War and obstructed the formation of a black middle class (Greason 2012). The *physical* spatialization that has structured the racialized layout of American cities and suburbs is matched by a *social* space that is segregated into a White Space and a Black Space with clear delimitations that come with associations of social place linked to skin colour (Anderson 2015; also see Simone 2016).[7] Moreover, while home ownership is often celebrated as a key integration avenue for new immigrant communities, America's 'political culture's celebration of the free market for homeownership' (Freund 2016) has contributed to a further segregation of African Americans who were restricted to the spaces that were deemed undesirable by whites and served with toxic mortgages. This type of discrimination was, again, systemic since 'throughout much of the twentieth century, discrimination by race was integral to the design, development, marketing and even financing of American cities and suburbs' (Freund 2016; see also Desmond 2017). Both African American exclusion and suburbanization are, as Dreier and Swanstrom suspected, structured and fuelled by public policies. One report in the aftermath of the killing of Michael Brown lists some of the critical ones:

> In St. Louis these governmental policies included zoning rules that classified white neighborhoods as residential and black neighborhoods as

commercial or industrial; segregated public housing projects that replaced integrated low-income areas; federal subsidies for suburban development conditioned on African American exclusion; federal and local requirements for and enforcement of, property deeds and neighborhood agreements that prohibited resale of white-owned property to, or occupancy by, African Americans; tax favoritism for private institutions that practiced segregation; municipal boundary lines designed to separate black neighborhoods from white ones and to deny necessary services to the former; real estate, insurance and banking regulators who tolerated and sometimes required racial segregation; and urban renewal plans whose purpose was to shift black populations from central cities like St. Louis to inner-ring suburbs like Ferguson. (Rothstein 2014: 2)

In Rothstein's view, the breadth of the effects of these and other public policies, like those regulating labour markets, won't allow continuing Ferguson 'as an isolated embarrassment' but will need to be seen as 'a reflection of the nation in which it is embedded' (Rothstein 2014: 2). The road from Lakewood to Ferguson is not linear. It is a network of suburban landscapes that now spans a continental imagination with global repercussions from Shanghai to Istanbul and Johannesburg. We will turn our attention to global suburbia in the following chapter.

Cosmo City, South Africa

6 Beyond the Picket Fence: Global Suburbia

The suburban has now become a ubiquitous and familiar landscape around the globe although it often still conjugates familiar tropes of the Anglo-Saxon suburb. Let's drive down the road out of Las Vegas in Donna Tartt's post-crash suburbs from her novel *The Goldfinch*:

> As we drove, the improbable skyline dwindled into a wilderness of parking lots and outlet malls, loop after faceless loop of shopping plazas, Circuit City, Toys'R'Us, supermarkets and drugstores, Open Twenty-Four Hours, no saying where it ended or began.... I looked up and saw that the strip malls had given way to an endless-seeming grid of small stucco homes. Despite the air of boxed, bleached sameness – row on row like stones on a cemetery – some of the houses were painted in festie pastels (mint green, rancho pink, milky desert blue) and there was something excitingly foreign about the sharp shadows, the spiked desert plants. Having grown up in the city, where there was never enough space, I was if anything pleasantly surprised. It would be something new to live in a house with a yard, even if the yard was only brown rocks and cactus.... The houses began to grow larger: with second stories, with cactus gardens, with fences and pools and multi-car garages.... A road had an imposing granite sign with copper letters The Ranchos at Canyon Shadows.... In Desatoya Ranch Estates, on 6219, where lumbers stacked in some of the yards and sand blew in the streets, we turned into the driveway of a large Spanish-looking house, or maybe it was Moorish, shuttered beige stucco with arched gables and a clay-tiled roof pitched at various startling angles. (Tartt 2013: 221–2).

This kind of landscape is now familiar to many around the world. Still, often, suburbs have rightly been discussed as a phenomenon associated with the Anglo-Saxon societal model (Forsyth 2012; Harris 2010; Jauhiainen 2013). Private land ownership, consumerist capitalism and the ideology of freedom prevalent in the United Kingdom and British settler societies have made Australia, Canada, the United States and to some degree Britain itself, ideal places for the prototypical single-family home residential suburb to thrive during the twentieth century. In fact, Lorenzo Veracini has interpreted 'both suburbia and settler colonial phenomena as premised on an anxious escape that comprehensively rejects environments that are perceived as increasingly threatening' and noted 'that suburbia *re-enacts* settlement' (2011: 340; emphasis in the original). Likewise, the suburban ideal is associated historically with nation building (Barraclough 2011). Relatedly, as Easterling has noted 'the developer William Levitt associated his suburban housing with familial and patriotic narratives that were particularly infectious in the post-war period and such stories accelerated the spatial effects of the house as multiplier' (2014: 90). The Anglo-Saxon single-family home suburb is often taken as the benchmark against which all suburbs are measured. In fact, as the eminent cultural and architectural historian and global city theorist Anthony King has observed, suburbanization (suburbia 'both as settlement form and lifestyle') is part of the more general narrative of modern urbanization that is written as the contrasting script to the 'traditional city' (King 2004: 97).

But Anglo-Saxon suburbia needs to be re-evaluated in a global context. To examine its origins and role in a global light, we start from the assumption that suburbanization precedes the modern period and capitalist property relations but attains particular significance as a form of the production of space in capitalist societies (King 2004; Walker 1981). It is necessary to depart from the common wisdom on suburbanization as a chiefly American domain. The Anglo-Saxon model, most often associated with the United States, is dominant in reality and in the literature and important to acknowledge. But it obscures historical parallels and alternatives in suburbanization elsewhere. There have always been different pathways to peripheral urban development, even in the American century that just ended. More importantly, in recent years, newer forms of suburbanization that give rise to the need for rethinking urban theory and research overall have sprung up around the world.

THE SUBURBAN PROSPECT: OBSERVATIONS ON GLOBAL SUBURBANIZATION

We need to engage in an exercise of provincializing the (North American) suburb (Lawhon et al. 2014). That said, the Anglo-Saxon 'picket-fence' tradition is neither insular nor the only game in town. The massive post-Levittown sprawl of single-family homes in North America – from New York to Lakewood, California and up through the peripheries of Quebec City to Ft. McMurray in the Alberta oilsands – has necessarily clouded our collective memory; we no longer keep in mind that modern suburbia was a hybrid product from the very start (Fortin 2015). The suburbs of the British Empire themselves had global roots. The modern American (or Anglo-Saxon) suburb certainly thrived on the colonial grid with its shared, property-oriented logic. This is where 'house lust', as Canadian urbanist Humphrey Carver called it, could unfold and be realized for the aspiring European immigrants.[1] But it got all its constitutive and spare parts from elsewhere. The bungalow, next to the veranda-fronted house, the ranch house and the single-storey villa, was an archetype of the single-family home not just in the suburbs of the United Kingdom, the Americas and Australia, but more generally around the world. A truly global phenomenon, the one-storey house has its origins in the distinctive imperialist history of Britain's colonization of India (King 2004). The imperial connection was kept alive as American consumerism was subsequently projected into the world with iconic images of suburban bliss; that America was ready to follow up the image with deeds was made clear whenever its armies put 'boots on the ground' to safeguard the empire's global interests in defense of the suburban dream at home (Keil 2007).

Still, the process of global suburbanization, even in the modern period, was never a mere extension of the North American model. We can identify three related lenses through which suburbanization can be viewed as a worldwide phenomenon:

1. **Suburbanization as modernization**
   Suburbanization is part of a larger process of the extension of systemic infrastructures and large-scale political economies. 'The house, its repetitive organization and the story attached to it all constitute information that contributes to disposition', writes Easterling (2014: 90). She defines disposition as 'the character or propensity of an organization that results from all its activity' (2014: 21). In that sense, suburbanization is part of the disposition of modernization, fraught with the contradictions of modern society's dialectics: 'The mass-produced suburbs sold unique country homes but delivered the virtually identical products of an assembly-line organization' (2014: 71). This kind of 'multiplication' might be accomplished through state sponsored, developer-built or self-constructed settlement patterns. The ordered segmentation of the self-built *favela* or *gecekondu*, the pattern of the peripheral high-rise public housing neighbourhood and the houses of Levittown display similar repetitiveness and modernization potential. They represent modernization as a disposition of 'an unfolding relationship between potentials' to return one more time to Easterling's formulation (2014: 72).

   The aspect of modernization includes producing more wide-ranging societal and spatial consequences, chief among them the emergence and division of modern social divisions such as class and race, the dynamic revolutions of industrialization and capitalist accumulation. Racialized and gendered class formation processes in particular are

written into the modern suburb and ride on its success as a constellation. Modernization in this sense is always also colonization.
2. **Suburbanization as modernism**
   While suburbs have existed since before the onset of industrialization and capitalist modernization as distinct periods, the mass phenomenon of building residential, productive, commercial and infrastructural spaces beyond the classical city's edge appeared simultaneously with the gesture of modernism, especially since the 1920s. This may be the reason why, as Fishman, in his quest 'to understand the significance of suburbia both for modern culture and for the modern city' observes, 'suburbia has always seemed contemporary' (1987: 4–5). Suburbia, to Fishman, appears as 'more than a collection of residential buildings; it expresses values so deeply embedded in bourgeois culture that it might also be called the bourgeois utopia' (Ibid. 4). Some authors think of this modernist gesture more as a signpost to an urban dystopia and make their more or less explicit critique of modernism (in sub/urban form, planning, design, etc.) a centrepiece of their evaluation of suburbanization (for Toronto see Sewell 1993; 2009; Kunstler 1993).

   Although some cities like Toronto are now surrounded by seas of conservative, gabled Victorianesque housing that signal more or less 'vulgar' subscriptions to family values and privatism, features generally observed in the landscapes of American neoliberal suburbanism (Knox 2008), suburbanization initially was the domain of modern and modernist reformers in architecture, planning and design. The Bauhaus style was predominantly a suburban intervention, be it through the individual 'master' houses or through the 'housing machines', high-rises and '*barres*', modern architecture put its stamp on the periphery everywhere. In the more benign and successful forms that have shown resilience to socio-economic and socio-demographic change, like those built during the programme *Das neue Frankfurt* under Ernst May and his colleagues or the – once peripheral – Karl-Marx-Hof in Vienna, the modernist promise continues to live; in the many 'habitats' of massive modernism on the peripheries of other European cities – east and west – modernism has lost its glamour and has become a symbol of social engineering failure (Logan forthcoming).

In the massive suburbs and cities of the world, modernist form and function are continuously revolutionized and recycled as recognizable pieces marked both by period (the early twentieth century) and timelessness, as contemporary in gesture. Lawrence Herzog says of Sao Paulo, for example: 'the new high-rise suburbs represented a compromise reached between modernist architects and the government. One contradiction, however, was that modernist architecture in its original form espoused egalitarianism, yet the new high-rise, modernist suburbs were mainly for the privileged, those who could afford to live in the new vertical tower compounds with their lush landscaping and heavy security.' (2015: 176–7).

3. **The centrifugal city**

For the classical suburb, separation, escape and segregation was the name of the game. Fishman's 'bourgeois utopia' is 'a partial paradise', a 'refuge' from the city's threats and threatening people (1987: 4). To Beauregard, building on Mumford, this is the 'parasitic urbanization' beyond the old urban form (2006: 41). The centrifugality is, of course, only partly literal, as an escape from the spatial centre. More importantly, it comes ultimately with an escape from society itself and as an involuntary process of peripheral segmentation and exclusion. Suburbanization, as Herzog says, once part of the promise of modern egalitarianism, becomes the vehicle – in both material and ideological ways – of splintering communities. Herzog notes for Sao Paulo that its vertical suburbanism comes with 'a disdain for public life' (2015: 175) and its high-rise neighbourhoods, often clustered in zones reminiscent of American edge city and similar to its horizontal American counterparts, are a place 'of escape from the perceived ills of crowded city life – crime, noise, discomfort' (2015: 175). Yet, it is not just the aspect of elite escapism that characterizes the emerging global suburbs; those are also where the poor end up that have nowhere else to go. To them, the remote suburbanism of the isolated tower neighbourhoods in the distant periphery become a trap. This phenomenon of exclusion and displacement associated with the centrifugal city is illustrated for example, in the case of Istanbul, where 'the urban poor find themselves relegated to high-rise apartments on the city's periphery, with no surrounding grounds and inadequate access to transportation, food and other basic necessities' (Dossick et al. 2012: 11). Yet, in reality, no real escape offers itself to these neighbourhoods as they are meshed profoundly and in many registers of dependencies into each other through mutual reciprocities of work, stuff, mobilities, etc. Standing on a fourteenth floor balcony of a newly built high rise in a secured upper-middle-class *condominio* in the Rio de Janeiro suburb of Barra de Tijuca, one can view the entire mix of the suburban assemblage in its diversity: to the northwest lies a favela nestled gently against the mountains and separated from Barra by a swamp and canals; just south of the favela is what used to be the athletes' village during the 2010 PanAm games hosted by Rio. The village, now condominium housing ties into the oversized transportation infrastructure that connects the area into the centre of Rio. To the east, towards the Atlantic Ocean, lies the bulk of the built up area of Barra, already well-developed with amenities of all kinds of retail and a large shopping mall on the opposing shore of an artificial lake. In the North, towards Rio, one sees more skylines of high-rises and the mirage of yet another athletes' village, that of the 2016 Olympics. This mixity of form, function and socio-demographics in the global suburb is also due to its innovative rearrangement of existing and emerging urban fragments that are condensed into a *bricolage* of novel proportions in the peripheries everywhere: in the West, immigrants bring practices and aesthetics that become part of the dominant vernacular; in the South and the emerging powerhouses of Brazil, Turkey, China or India, pieces of postcolonial globalities are glued together into new landscapes of formal and informal sub/urbanity. In those global suburbs you find, then, what Simone (2012: 40) with reference to Nigel Thrift has called 'fugitive materials' – traditions, codes, linguistic bits, jettisoned and patchwork economies that are 'on the run', pirated technologies, bits and pieces of symbols floating around detached from the original places they may have come from.

THE REGIONS OF SUBURBANIZATION

Outside and beyond the American case, there is a spectrum of more or less endogenous developments in suburban form and process. While many suburban developments around the world mimic the American example as Herzog has demonstrated for Brazil, for example, where 'copying the United States has been a conscious part of the marketing strategy of Brazilian developers' (2015: 175), suburbanization has been as much a product of situated local governance and development dynamics.

Canadian suburbanization fits the Anglo-Saxon tradition in many ways. Canada can now be called by all statistical measures a 'suburban nation' (Gordon and Janzen 2013). Still, its trajectory is markedly different from the American one. Before the Second World War Canadian cities were a more fragmented, diverse patchwork shaped by de-centralized industry stretching along rail corridors and social space segmented along class, ethnic/racial and religious divisions (Addie and Keil 2015). Regional diversity in pathways towards becoming a suburban nation was remarkable in the continent-spanning expanse of the country (McCann 2006). One important aspect of settlement in Canada was self-built working-class suburbanization in unserviced fringe areas of the kind Richard Harris documented in his magisterial *Unplanned*

Suburbs (1996) for the Toronto case. The resulting neighbourhoods were significantly different from the traditional bourgeois-utopian model. Harris also exposes the history of how the country's idiosyncratic self-built suburbanity increasingly made way, after 1945, for a 'creeping conformity' of corporate mass-constructed housing in planned subdivisions (R. Harris 2004). The pervasive conventional suburbanization was made possible by significant state intervention, especially subsidized mortgage financing and infrastructure investment (Keil, Hamel, Chou, Williams 2015). Yet, Canada, in breaking with the tradition of its neighbour to the south, of single-family home suburbanization on peripheral green fields, also witnessed the building of large-scale housing neighbourhoods on its cities' peripheries in a style reminiscent of the French *banlieue*, German *Großwohnsiedlungen* or *Plattenbausiedlungen* and British tower estates. Especially in Toronto, where more than one thousand high-rise towers were built after the 1960s, most of them in the periphery, a peculiar landscape of inner suburbanity arose (Charmes and Keil 2015; Young and Keil 2014).

In Europe, a broad spectrum of suburban forms and processes developed throughout the twentieth century (Phelps 2017). Perhaps the first most visible alternative to the picket fences of North America are the large housing estates that were built on the 'green' periphery all over the large Fordist North Atlantic countries such as Great Britain, France, West Germany and Canada. In contrast to the mostly privatized suburbanization in North America (albeit subsidized through tax and other incentives by the state), suburban housing during high Fordism in Europe was mostly produced by state-led or co-op development and management corporations. In both Western and Eastern Europe, these estates became the focal points of a new urban crisis at the beginning of the twenty-first century, as planners and developers started to cultivate buzzwords such as density, compactness, 're-urbanization', New Urbanism and 'the creative city' (Charmes and Keil 2015). These estates arose on an even larger scale in the countries of what friends and foes of the Warsaw Pact states came to call 'real existing socialism' – the former Soviet Union and its satellites – where hundreds of millions of housing units (and factories) mushroomed in peripheral urban extensions. European suburbs are characterized by a large 'variety apparent in their form and modalities of governance and indeed the mixing of these forms and modes in individual national settings' (Phelps and Vento 2015, 155). A recent book on the subject of European suburbanization recognizes that the continent's peripheral urban developments are both old and new. Phelps writes:

> There are multiple senses in which European (sub)urbanization might be thought of as 'old'... First, processes of urbanization have proceeded earlier and further in many parts of Europe than elsewhere.... Second, Europe possesses the oldest nation states of this modern world system and the most established frameworks of state territoriality within which the process of suburbanization takes place. Finally,... the novelty of processes of sub(urbanization) in the US are often presented in some contrast to those prevailing in Europe. (Phelps 2017)

Among those 'old' characteristics are the 'non-contiguous, scattered and piecemeal' form, 'involving urban extensions tacked onto towns and cities' with wide 'variations on this theme of discontinuous developments at the periphery promoting polycentricity alongside economically important historic cores' (Phelps and Vento 2015: 158). In addition, even in countries with Anglo-Saxon heritage such as Ireland and Britain the garden city and cottage-style suburbanization as well as the odd working-class suburb (often in mining areas) was joined by large-scale, state-sponsored or even -built, suburban housing neighbourhoods such as Ballymun in Dublin (Farrell and Kelly 2015) or the poorly served and isolated English tower estates at the peripheries of London, Birmingham or Glasgow, resulting in 'a curious suburban landscape in which towering office blocks of the central area loom over the surrounding suburban detached housing' (Mace, Phelps and Jodieri 2017; Phelps and Vento 2015). Yet Britain remains understood more tightly as part of the Atlantic experience. To see real difference also in conceptualizing suburban developments at variance from the American case, one needs to look to the European continent.

The French case is instructive and well documented (Charmes 2005; Epstein 2013; Rousseau 2015; Dikeç 2007). Lefebvre reminds us of the explosive growth of the Paris suburbs after the Second World War: 'You know that there were very few suburbs in Paris; there were some, but very few. And then suddenly the whole area was filled, covered with little houses, with new cities' (Lefebvre and Ross 2015: 50). The conditions under which these post-war suburbs and their successors were produced (and consumed) differed greatly from the Anglo-Saxon prototype. For one, there was always more direct involvement by the state (Dikeç 2007; Kipfer, n. d.). The building typologies were mixed as well. The 'little houses', the pavillons that Lefebvre saw originally mushroom around Paris grew in tandem with the *grands ensembles* (towers in the park, disparagingly called the 'habitat' by Lefebvre). Both morphologies congealed into a specific post-suburban landscape with variable densities that now house much, if not most, of the metropolitan population in France's large urban centres (Charmes and Keil 2015).

While the German *Vorstadt* is a faithful translation of suburb into German, this term denotes a variety of morphologies from the free-standing bungalow on a large lot through row houses and mid-rise apartments to Corbusier style tower estates. In early twentieth-century Germany, suburbanization was often the stated goal of modernist reformers who advocated (and built) satellite developments on or outside the *Stadtkante*, the edge of the city. Symptomatic of this development in the early twentieth century was May's *Das Neue Frankfurt* mentioned above. After the Second World War, the often polycentric West German industrial landscapes and their rural hinterlands were dotted with numerous but scattered single-family home peripheries, punctured by mid or high-rise developments predominantly built in the modality of co-operative or social housing. Perhaps the most striking form of suburbanization in West Germany was the ubiquitous *Großwohnsiedlung*, the large housing estate, which collared the classical urban cores from Berlin, Märkisches Viertel in the East to the Neue Vahr in Bremen in the North, Bonames in Frankfurt in the centre and Freiberg in Stuttgart in the South. The post Second World War planners and architects who were responsible for the construction of millions of high-rise housing units in the West German periphery were largely driven by the same modernizing zeal as their predecessors in the 1920s (and strangely unaffected by the reactionary, traditionalizing Nazi period). They were part of a global modernist reform movement by urban professionals for whom the suburbs and Bauhaus modernism were ideal partners in striving for a sub/urban form that coincided with a general socialist or humanitarian egalitarianism in general or social democratic welfare state rationality in particular. Like in their British and French counterparts, those suburban tower estates became symbols of failure and decline in the decades following the 1980s. But they have also become integrated pieces in a polycentric in-between landscape of multiple suburbanisms like in the Ruhr or Rhein-Main areas (Basten 2016).

The West German view of crisis in the large-scale housing estates was compounded and ratcheted up a few notches in the treatment of the East German periphery, especially after 1990. While in the eyes of and so intended by the socialist planners of the post Second World War years in East Germany and elsewhere, in the Warsaw Pact states large-scale pre-fabricated housing was built as urban extension rather than suburbanization, many if not most of the massive estates built in the GDR and elsewhere in the East were in fact suburban in location and displayed some of the same formal and functional characteristics as their suburban counterparts in the West. While often larger in scale, they did follow the separation of land uses common to the Charter of

Athens era of European planning. Well-connected to regional public transit, they tended to be tied into an oversized public road system of highway and arterial roads (where admittedly the density of automobile traffic was by far not as high as in the West). After 1990, many of these *Plattenbau*-estates fell into disrepair and crisis as they became objects of privatization and corporate investment and symbols of everything the West considered wrong about socialist planned economies. Only in the past decade, often following the painstaking work by urban scholars and activists, have the peripheral housing estates in East Germany begun to be rehabilitated as valuable, sustainable and renewable sub/urban housing alternatives (Bernt 2009; Kabisch and Rink 2015).

Since the fall of the Berlin Wall, 'post-socialist' Eastern Europe, more generally, has been characterized by liberalization and deregulation of urban planning which has led to sprawl beyond the traditional peripheral high-rise estates (Szirmai 2011). Stanilov and Sykora (2014b: 316) note: 'Suburban sprawl, which was virtually unkown before the fall of the communist regime, has become a defining feature of the postsocialist metropolitan landscapes, where numerous fragmented clusters of new residential and commercial developments are spreading far beyond the boundaries of the compact urban cores.' The same authors bemoan that in the region '[t]he explosive growth of the suburbs has impacted negatively the ability of central cities to regenerate some of their areas threatened by long-lasting urban decline' and that vast suburban developments have weakened the strategic ability of the state to deal with the social and spatial contradictions of de-industrialization and inner-city decline or shrinkage of its massive housing estates (Sykora and Stanilov 2014: 21). In addition, the rapidly privatized Eastern European periphery has become the site of various degrees of gating. It is here where the global tendency of the centrifugal city is on full display and develops some of its most bizarre constitutive forms (Hirt 2012). The socio-spatial inequalities and contradictions between shrinking socialist housing estates and fancy gated single-family home estates abound in cities from Belgrade to Sofia, from Tallinn to Leipzig. The Eastern European case is not just of importance because of its unique trajectory under post-Socialism. It has also generated a particular and interesting theoretical perspective on suburbanization (Stanilov and Sykora 2014a).

GLOBALIZED SUB/URBANIZATION

The second most prominent alternative are the ever-expanding self-built and squatter settlements in urban areas of the Global South – that is, Africa, Latin America and developing Asia, including the Middle East.

What is true for the increasingly inappropriate and impossible use of Northern models to make theoretical sense of global urbanization patterns, is particularly visible in the suburban areas of the globe's emerging urban fields. Surely, the classical North American case remains relevant to our understanding of the history and geography of suburbanization and its theoretical implications. Not least Los Angeles is forever marked as the unsuccessful sister city to hyper-dense New York (Swilling 2016). Yet, perhaps contemporary squatter settlements of Cape Town's Mitchell's Plain more than Los Angeles's post-war middle-class community of Lakewood is the true icon of contemporary suburbanization. In Asia (McGee 2013), Africa (Mabin, 2013) and Latin America, 'the post-colonial suburb' (Roy, 2015) is taking shape as a global form and it points back to those suburbs of the North, be they in Europe (France, the UK, Germany, Portugal, etc.) or Canada, where large numbers of immigrant communities have begun to settle on the fringes of the cities. King speaks of diversity in the age of the postcolonial 'globurb' by which he means 'forms and settlements on the outskirts of the city, the origins of which – economic, social, cultural, architectural – are generated less by developments inside the city, or even inside the country and more by external forces beyond its boundaries' and adds importantly that

[d]ependent on their location, many of such suprurbs or globurbs, as with previous historical experience, continue to be generated not just by 'international' or 'global' forces but, more particularly, by those of imperialism, colonialism, nationalism, as well as the diasporic migratory cultures and capital flows of global capitalism. (King 2004: 103)

In the Global South, urbanization exploded predominantly as a result of the country-to-city migration that began in the middle of the twentieth century, first in Latin America and more recently across Asia and Africa. The majority of the newcomers move to places on informal urban peripheries, often squatter settlements, resulting (as in São Paulo) in 'a dispersed city in which long distances separate the center from the peripheries' (Caldeira 2013). In the metropolitan regions of Turkey, China and India, large-scale high-rise neighbourhoods surround the urban centres where many of the in-migrants and those expelled from the gentrifying inner cities live. A large part of this primary peri-urbanization – this creation of the urban fringe – materializes in the form of gated communities.

Importantly, models of peri-urban expansion strongly developed in several regions of the non-Western world. Most famously, perhaps, the (originally Indonesian) notion of *desakota*, first identified with Terry McGee's (1991) work and meaning the places inbetween traditionally urban and rural areas. More than just a geographical concept of peri-urban location, the *desakota* refer to interlinked socio-economic relationships between city and countryside (Moench and Gyawali 2008). The concept *desakota* has more recently been used as a launching pad for in-depth empirical studies of urban fringe developments in Asia. Ortega (2012: 1119), for example, in his work on Metro Manila, has investigated 'a more complex process of rural-urban change, moving beyond a mere consideration of in situ integration of agricultural communities into urban economies' and has identified a variety of spatial developments and argued for 'a place-based articulation of neoliberal restructuring.'

Based on Terry McGee's pioneering work, Ortega further points to the Western bias in urban theory building that has led to teleological and false universalist understandings of urban and suburban development everywhere. In response to this experience, Ortega proposes to use 'desakota as a hinge concept' and invokes 'its original anti-essentialist, "pagan", resistant and open character to articulate a production of space that is grounded on empirical understandings of relations and processes' (2012: 1122). In Ortega's work, the urban periphery in South East Asia is not a place outside of urban history and geography. It is a place from which theoretical claims can well be made. It is also not a place that is marked as marginal geographically and placed in an imagined bygone period of history but very much a product and producer of variegated neoliberal sub/urbanization processes today. In fact, as Hudalah et al (2016) have demonstrated looking at peri-urban gentrification in Indonesia, the fringes of the city in that part of the world are a productive place of both aggressive urban change and seedbed for new and relevant understandings of key concepts in urban studies. Similar claims are made in work on Mexico City by Aguilar, Ward and Smith (2003: 4) who have studied the 'regional "penumbra" of mega-city development – i.e. the peri-urban region – rather than just focusing upon suburban expansion or urban restructuring'. In doing so, they widen the empirical and conceptual terrain from which such developments can contribute to our understanding of extended urbanization globally (see also Gilbert and De Jong 2015).

Suburban cityscapes are now so ubiquitous that they have led to a re-evaluation of urban form and process overall, throwing into sharp relief the multi-dissected, no-longer-central structure of suburbanized landscapes. We have entered a state of continuous 'post-suburbanization' – a term mostly trying to denote that historical suburbs are now in a complex, politically contentious, renewal and urbanization process. In the introduction of this book, we posited the four dimensions of urbanism following Roy's insightful essay on the subject. As we take the 'marked' American suburban model 'global', we need to account for the baggage that comes with the terminology we employ. If suburbanism is global, it will be important to remember the conventional implications of this term:

> Urbanism... has come to refer to a distinct kind of site (the city), separable from other rural places, and taken to be a hallmark of modernism, progress, development, and the metropole – the opposite of provincialism. At the same time, urbanism is associated with a set of social ills, the dark side of development contrasted with an idyllic rural past. This dissonance implies the need for intervention – urban planning to achieve development while minimizing a social dysfunctionality. (Sheppard et al. 2013: 893–4)

In the Western, normative discourse on urbanism, this normally also implies that the practice 'takes for granted that capitalism and liberal democracy are natural, ubiquitous norms and capable of over coming the poverty, inequality and injustice seen as so pervasive across the global South' (Sheppard et al. 2013: 894).

Similar discomfort is expressed in Ash Amin's essay on 'telescopic urbanism' (2013) that he confronts with demands for a politics of the urban staples couched in a language of city-wide infrastructural rights. Telescopic urbanism of the kind he critiques make invisible 'the myriad hidden connections and relational doings that hold together the contemporary city as an assemblage of many types of spatial formation, from economically interdependent neighbourhoods to infrastructures, flows and organisational arrangements that course through and beyond the city' (2013: 484). Among these assemblages are the 'disjunct fragments' in Lefebvre's imaginary, yet they have no linear connection back to the centre in this scenario. Suburban forms and lives are not necessarily related back to a singular centrality. The important point is to integrate them into our general understanding and theorizing of urbanism.

Suburbanization was considered part of the American dream. And so has been its crisis. From Brazil to India, from South Africa to China, American suburbanization provides an important model for the global urban middle classes and elites (Hamel and Keil, 2015; Herzog 2015; Keil 2007). But the imagination of a globally scaled and themed suburbanity now comes from elsewhere, not from the American heartland. As Roy has reminded us, non-central worlding practices now include strategic suburbanizations as part of a multilogue of inter-referencing in the Global South itself (Roy 2009; 2015). In joint work with Aihwa Ong, Roy notes: 'While such practices can be read in the register of urbanism, it is important to make note of their worlding character' (Roy and Ong 2011) including endogenous forms of neoliberalization.

The significance of the global suburban in this context is not just a broadening of the empirical base into colourful multiplicities, it is a more far-reaching claim towards the rewriting of suburban theory. It needs to take to heart what Simone said about studying cities: 'As for urban theory then, understanding the systematicity of the city entails thinking of cities as many different cities at the same time, not as a plurality of fractals, but as the designs and struggles of many attempting to recognize each other as one, always imposing themselves on the other, as well as finding ways to leave each other alone' (2012: 46). Understanding the systematicity of the global suburb demands the same. Seeing suburbanization in that mode provides the grounds for a new type of theorizing that retains a critical, materialist edge yet demands the rethinking of the multiplicity of suburban form and function from the periphery. American suburbia (and by extension the European high-rise periphery of the Fordist and socialist period) are removed from their roles as conceptual and policy models for global suburbia. Theories of and on suburbanization are simultaneously rewritten from those new (real and imagined, material and intellectual) locales of the production of suburban space that mushroom between Johannesburg and Moscow, Shanghai and Santiago (Robinson 2016).

Amadora, Portugal

The global post-suburb is now a layered sediment of contradictory temporalities and spatialities, an assemblage of colonial and postcolonial histories and geographies. This takes wildly different forms. Let us consider two different European cases, one with a decidedly postcolonial history, Amadora, in the north of Portugal's capital Lisbon and the other post-socialist, the sprawling periphery of Belgrade in Serbia. Both are complex palimpsests of tumultuous histories, formal and informal settlement, state and market intervention (benign and punitive) and widespread self-built houses, marginal economies, blossoming gardens and damaged ecologies. If we look outside of Europe, we have as one example the Delhi peri-urb of Gurgaon, described by Gururani as one example of a larger phenomenon of urbanization in India overall:

> Located at the fringes of metropolitan cities, urban peripheries represent a frenzied urbanizing frontier, a rural-urban interface that is typically

characterized by mixed land-use, intense development and fragmented pockets of wealth and deprivation. Clearly, like in much of the Global South, in India, the process of peri- or suburbanization is as much about agrarian change as it is about urbanization. In most scenarios, it tends to entail a highly volatile, even violent, process of land acquisition, displacement and development. (2013: 183)

China experiences similar dynamics of building new urban worlds in the periphery of large historical urban centres such as Beijing and Shanghai; it has produced massive suburban and new town developments, in a mix of primary land market formation (as rural communal property is transformed into privately owned housing units and corporate factory sites) and state urban development policy. The outcome is an architecturally

Belgrade, Serbia

polyglot periphery where one finds the design of a midsize German industrial city next to that of an English market town (Wu and Shen 2015), all overshadowed by forests of gigantic high-rises and connected through high-volume road and transit infrastructure to each other and to the city centre (Fleischer 2010). These new towns are, in contrast to some of their American or European models and counterparts, not preferred to inner-city locations by a discerning and newly empowered, socially mobile clientele seeking work or shelter: 'In fact, most urban residents tend to avoid moving to the suburbs if they have a choice, as Chinese suburbs are not yet fully developed places with adequate basic amenities and shopping, schools and social activities are still lagging behind those in the city' (Pow 2012: 58).

In conclusion, we can note that the trajectory of suburbanization from the American real and imagined origins to global ubiquity is now in clear view. However, we also know that what we find at the urban peripheries around the world does not currently, and has rarely done so historically, follow a particular blueprint that comes from the West. This chapter traces through its regional rationality a few select cases of suburbanization that sometimes precede its alleged Western models. Suburbanization, then, appears as always constructed through different circumstances. Still, the regional specificities must be understood as emerging in connection with broader developments such as globalization and neoliberalization. Regionalism or even localism has to be overcome conceptually through critical post-suburban theory that remains open to innovation. New geographies of theory must not be leading to new reifications of perspectives. What we have learned from examples taken from Belgrade to Beijing, Rio to Istanbul, Mexico to Manila, Istanbul to Lisbon, is that the road from Los Angeles leads down many different pathways of suburbanization and suburban theory.

Subway Station, York University, Toronto

7 Suburban Infrastructures

Suburban infrastructures is a subject area of considerable interest in sub/urban studies. Extant work on suburban infrastructure includes conceptual and historical considerations (Addie 2016, 2017; Filion 2013a, b; Filion and Keil 2016; Filion and Pulver forthcoming), concerns about vulnerabilities (Bloch, Papachristodoulou and Brown 2013; Monstadt and Schramm 2013; Young et al. 2011), the role of infrastructures in regional governance (Addie and Keil 2015) and the significance of infrastructure in a new post-suburban politics (Young and Keil 2014). Building on these lines of thought, this chapter makes a simple point: Suburban areas perform vital infrastructural functions in the metropolitan region and beyond. The majority of analyses on infrastructures in suburbs emphasize their role in the building of housing subdivisions and other suburban developments such as industrial parks (roads, ring roads, water and sewage, etc.) and the infrastructural deficit often experienced when cities grow fast at the metropolitan periphery. Also, importantly, suburban infrastructures, especially the entangled landscapes that followed the spreading of post Second World War highway networks, are complicit in the destruction and segregation of working class and people of colour neighbourhoods and the basis of building (white) suburban privilege into the fabric of the urban region (Baum-Snow 2007; Grunwald 2015; Poitras 2011; Stromberg 2016). Debates on urban sustainability in an age of climate change note the automobile dependency of suburban infrastructure arrangements and contrast those with normative preferences for dense, walkable and transit-oriented developments. Given the legacy of sprawl fuelled by cheap oil and ravenous appetite for profit by the suburban growth machine (Logan and Molotch 1987) and the expected ongoing global suburbanization in an exploding urban world (Swilling 2016), the significance of infrastructure in metropolitan peripheries for human life, economic growth and ecological sustainability cannot be overstated.

While these are important points and concerns, in this chapter I focus instead on the infrastructural functions that are located in and emanate from the suburbs and that provide services there and beyond. An important, underrated aspect of suburban infrastructures is their tremendous importance for the functioning of the entire urban region.[1] While we often hear of the relocation of certain prime functions into 'back offices' in the urban periphery and while this is thought of as a hierarchy caused by land markets, the debate on suburban land use tends to underplay the necessary and valuable services the urban periphery delivers to the urban region in general and the urban core in particular. This includes prime network spaces such as airports, recreational spaces such as golf courses, but also noxious and toxic uses such as garbage dumps, water treatment facilities, incinerators, etc. These infrastructural arrangements between the city and the suburb (in which the suburb mostly takes the subordinate role) include biophysical and metabolic relationships as well as technical connections of all sorts: from farmers' fields (for local food diets) to sewage and compost recycling and high tech agriculture in hothouses, the suburbs are part of the metabolic matrix from which the city nourishes itself both as a mechanical and corporeal entity (Hume 2015). The flip side of these 'infrastructural services' provided by suburbia, is the massive resource expense that goes into these networks, hubs and sites that structure suburban space. Anthony King reminds us of 'the immense proportion of a society's resources invested in' the suburb, an effort equivalent, as he notes, based on Taylor, to 'the great Gothic cathedrals of the high Middle Ages of feudal Europe' (King 2004: 98).

Suburban infrastructure, often thought of as merely functional for the suburban constellation itself, turns out to be multi-scalar and supportive of metropolitan and higher-scale purposes. Avoidance strategies put in place by regional infrastructure planners support suburban lifestyles. This is the logic of the bypass that has taken over the suburbanized region. Infrastructures in this context also work as a sorting mechanism of complex suburban landscapes. In this context, suburban transportation infrastructures, for example, are playing a significant part in the location and delivery of the metropolitan mobility structures overall. Among all infrastructure issues, mobility stands out. The chapter will discuss some of the important issues related to transit, intrametropolitan goods transportation, management of automobility but also larger scale issues connected to the suburban fabric such as warehousing, air travel, regional water, wastewater and waste management and so forth. Growing mobility changes the sub/urban landscape which appears without boundaries; movement and connections become uncharted and unmoored (Bertolini 2012). Instead of just being seen as a burden (automobile-dependency),[2] suburban infrastructures are coming into relief as a segment of a system that serves multiple needs of people who 'live in one place, work in a second and shop, care for another person, or seek entertainment in another' (Bertolini 2012: 16). Equally, '[b]usiness processes are also becoming more and more spatially articulated'. Both tendencies provide more freedom to users as mobility increases but that also makes mobility more of a constraint and necessity. The complexity of user demand and the insufficiency and inappropriateness of supply combine to make the traditionally predictable transport land-use/feedback cycle a source of much uncertainty. What is worse is that such uncertainty is 'irreducible' which calls for new planning and policy strategies to provide acceptable mobility across the suburbanizing region (Bertolini 2012: 17).

The governance of mobility in the suburbanizing region encounters growing dilemmas and difficulties along various lines. Practitioners and theorists have reacted to these challenges by discarding the former 'predict and provide' and 'predict and prevent' approaches in favour of a paradigm of 'sustainable mobility' that tries to 'identify forms of mobility which acknowledge the need and desirability of mobility and, at the same time, can reduce its negative effects' (Bertolini 2012: 18). Still major problems persist: comprehensive infrastructure projects are hard to install in a region that is divided between 'urban' and 'suburban' interests or fragmented and splintered into incompatible or worse, contradictory relationships of uses; cost is not accepted by an electorate suspicious about long-term state deficits and debts (meaning taxpayer liabilities); technological change is too rapid to make sound decisions that can be defended publicly. Thus, we often end up with 'congestion for those who drive or ride and exclusion for those who do not' (Bertolini 2012: 18). Achieving transit equity is difficult in a post-suburban region with large deficits in one corner and oversupply in another, with 'both too much and too little' (Young et al. 2011) infrastructure that bypasses poor and immobile populations (Hertel et al. 2015, 2016) and in a political system that structurally works against the in-between spaces that need upgrading most (Young and Keil 2014). Given the gravity and scale of the issues of mobility in the 'real existing' suburbanizing region, decision making needs to take place with multiple stakeholders at various scales involved; those stakeholders must reconcile very different interests of mobility in force field of often contradictory territorial interests, competing and fast changing technological solutions and contested political discourse (Addie and Keil 2015; Savini 2013).

In this chapter it is my intention to move the debate from 'the infrastructure is the message' (Filion 2013a) to suburban infrastructures as lifelines of the urban region. Filion has noted perceptively about suburban infrastructure: 'Infrastructures can be seen as the media and their message as the form suburban development takes. In other words, the contribution of infrastructures is not limited to their assigned task of conveying flows. Accordingly, transportation and sewage networks do not only carry people and goods or evacuate human waste; they also have a determining effect on the location and form of urban development, especially on its density. The same goes for other infrastructures: water distribution, electricity and electronic transmission, for example' (2013a: 40; see also Hesse 2013). The focus of this observation remains the suburb itself that is produced, shaped and served by the reach, scale and mode of infrastructure at its base. Not just suburban form, also suburban life is shaped in this manner: 'Time budgets and work and consumption behaviour are tributary of the nature of activities present in suburbs and their distribution. So, as infrastructures are a major factor in the shape taken by suburban development, they de facto contribute to the suburban lifestyle' (Filion 2013a: 40). It is important to note that neither Filion nor Marshall McLuhan from which he draws his inspiration are technological or infrastructural determinists but see the effects of technology on (suburban) form and life as mediated through social and political decision-making processes. It also needs to be added that the ways in which form and life intersect in suburban environments has much to do with the systemic conditions under which they exist: suburban infrastructures may have led to very similar looking morphologies, for example, in East or West German peripheral tower housing estates (*Grosswohnsiedlungen*) but life in those complexes differed in significant ways before 1989; similarly, mobility infrastructures may lead to similar looking transit oriented developments in Chinese new towns and the periphery of Dublin but life in those settlement areas is characterized by very different accessibility to labour markets, commercial facilities, social welfare services, etc. Lastly, suburban single-family home neighbourhoods in both Europe and North

America may owe their layout to automobile dependency but the likelihood of having a cycling infrastructure linking residents to shops, schools, services and transit is much higher in Europe than in the US or Canada. Not all infrastructure 'messages', therefore, can be read in the same way.

Infrastructures can be understood as one of the foundations of building and understanding the urban periphery. Suburban infrastructures have become the lynchpin of urban mobility and circulation, information distribution and socio-natural metabolisms (Keil and Young 2009). Although much attention is often paid to the 'last mile' of infrastructural bundling in downtowns of global cities (Graham and Marvin 2001), the suburban grids, their functioning and deficits, have become a major focus for both soft and hard infrastructures. Physical infrastructures such as mobility, water and sewage, communications and energy and social infrastructures such as those for health and education, are crucial dimensions of suburban development (Filion and Keil 2016). In 'modern' cities and suburbs, these infrastructures are assumed to be networked and connected both upstream and downstream to different scales through which they are established and realized: from global economic and financing or technology development to the local bus stop, the faucet in a suburban home and the cell phone tower at the corner, infrastructures articulate complex, large technical systems (Hughes 1987) that establish life in sub/urban environments. In reality, networking as part of the 'modern infrastructural ideal' may only be achieved in a fraction of the world's cities and even less so in the peri-urbanity of the global South or East. Once the symbol of urban integration, infrastructures in the West may indeed 'splinter' metropolitan regions confronting suburbs with higher costs and/or lower quality of services. In the postcolonial world, where the universality of suburban infrastructure provision was never realized, new innovative infrastructure technologies and governance modalities are associated with suburban growth (Kooy and Bakker 2008; McFarlane et al. 2014; McFarlane and Rutherford 2008). In many areas of the globe where today we find the most dynamic and aggressive forms of sub- or periurbanization, it is not *physical* networking but *social* networking that makes the difference, a process that Simone has in mind when he speaks of 'people as infrastructures' (Simone 2004). It is not unusual to see modern, networked infrastructures existing alongside the improvised, constantly and performed mobilities and connectivities of the people in the rapidly exploding peripheries of today's metropolitan areas (Siemiatycki 2006). In many instances, previously assumed trajectories of development from local, unconnected provision of a particular service (say, water from a well) to higher and more complex orders of networked articulation (say, to a reticulated, piped municipal water system) cannot be taken for granted: structural inequities lead to parallel pathways of development (Gandy 2014; Swyngedouw 2004) and progression of technological evolution previously presumed necessary may be sidestepped and bypassed (as has been the case for the distribution of cell phone networks instead of land lines in many of today's suburban environments). These growing discontinuities in the technological and organizational emergence of infrastructures also calls into question the traditional divide of formal and informal infrastructures that was considered to be a firm marker of modernity in the twentieth century. Today's suburbanization relies on both and it is the rich as well as the poor who are dependent on various shades of formality and informality in the delivery of the services necessary to sustain sub/urban lives and economies. Infrastructures are therefore both expressions of existing and avenues for emerging political, social and economic power and tied into the governance of suburbanizing regions (Hamel and Keil 2015) and the development of land (Harris and Lehrer 2018). They are also keystones of sub/urban metabolisms, especially in the establishment and definition of boundaries and transitions of the global suburban with the natural environment, blue-green landscapes and post-suburban political ecologies (Monte-Mor 2014a, b). Dynamic and emerging societal relationships with the natural environment and new urban political ecologies signify an important aspect of suburbanization around the globe (see chapter 8).

Some of the suburbanization of infrastructure is part of the process of making networks of urban metabolisms invisible, especially in the case of regional transit and transportation infrastructures (Kaika and Swyngedouw 2000). It is, in fact, part of the splintering of cities into – differentially – functional and valued spaces in the urban region (Addie 2016; Graham and Marvin 2001). The hard and soft infrastructures of the flexibilized post-Fordist global economy that settle in the suburban rings of metropolitan areas (around airports, in business parks, malls, etc.) have become vital segments of a globalized network of tracks, lines, microwave towers, runways and highways as well as relay stations of a globalized metabolism of the urban revolution.

In all these ways, suburban infrastructures are central to the globalized economies and regionalized lifeworlds of massive populations as they operate as conduits and facilitators of economic and everyday life. They can be the central terrain of suburban extension and important pathways of both implosion and explosion in the sense of Lefebvre's formulation (2003). Suburbanism is a lifestyle with an extended landscape attached. This landscape is ordered and made accessible by infrastructure. The suburban lifestyle stretches across vast landscapes of extended urbanization. The suburban infrastructure problematique is linked directly to the governance of suburbanization through state, accumulation and private authoritarian means (Hamel and Keil 2015) as well as the production of suburban land (Lehrer, Harris and Bloch 2015).

Suburbanism systemically extends its territory through the proliferation of technical infrastructures such as transportation networks and water/sewer grids. But there are also, especially in the non-conventional, informal variants of suburbanization that have much impact globally today, so-called 'infrastructure deserts' in both emerging and existing suburban landscapes. These deserts can take multiple forms either in absolute terms, such as the total lack of access to fresh water in a squatter settlement or in relative terms, such as being shut out from, or 'bypassed' by, the prime infrastructure networks and having to rely on second rate, slower (but often equally or more expensive) means of mobility while more privileged system users have access to first class, faster or even cheaper means of transportation (Filion and Keil 2016). In other contexts, informal infrastructures provide relevant lifelines in urban and suburban contexts as is the case with small vehicle, small-scale operators in cities in the Global South that 'allow car-less, disadvantaged individuals to reach jobs, buy and sell produce and access medical care' (Cervero and Golub 2007: 456). Significantly and in line with peripheral urban development, these modes of informal transportation 'also enlarge laboursheds, expanding the supply of workers across many skill levels from which firms and factories can draw upon' (Ibid.). Those developing cities while sprawling at the periphery 'have a more monocentric urban form than their developed-city counterparts. In many African and South American cities, for example, a third or more of formal jobs are concentrated in the urban core, considerably above what is found in most US and European metropolitan areas' which makes infrastructure innovation necessary and possible (Cervero 2013: 9). And even where more formal modes of transportation are part of the peripheral expansion as is the case in most Chinese new towns, there are severe issues around access, cost and travel times (Cervero and Day 2010).

Some poststructural commentators have put forth the suggestion that infrastructure spaces (and suburbs) as actor networks of sorts can possess 'disposition just as does the ball at the top of an incline (…) Disposition is immanent, not in the moving parts, but in the relationships between the components' (Easterling 2014: 72; 2010). This space is 'thick with technologies that are potential multipliers: populations of suburban houses, skyscrapers, vehicles, spatial products, zones, mobile phones, or global standards' (2014: 217). In this sense and following Easterling further as she speaks about the vast conurbations emerging in China, the suburbs are a 'zone', suburbanization is a horizontal division of labour, a giant production grid, a gargantuan spatial factory floor spread across city and society – enabled by networked infrastructures mostly.

Suburban infrastructure has to be understood, at least in part, as a mediator and result of capital investment into the metabolism of metropolitan areas. Sold as an infrastructural landscape to serve private needs, suburbia is also a substantial fraction of fixed capital that 'works' at the behest of larger accumulation processes and interests. This role of suburban infrastructures as the fixed capital backbone of current production-of-space induced accumulation is often hidden from plain view. Like a true fetish, the vulgar monstrosity of the single-family home with the two or three car garage that is at the end of the infrastructural grid is mistaken as the cause of the problem of the metropolitan crisis today instead of seeing it (and its inhabitants) as products and addicts of a larger systemic infrastructure of production and consumption that serves the larger goal of capital accumulation. Use value – 'the house' – is foregrounded and exposed where exchange value – 'reproduction of labour' – is the real objective. Measuring infrastructure needs from that angle will cast a new light on the debate.

In fact, when it comes to global suburbanization, considerations regarding infrastructures play a particularly important role. In the epilogue to a state of the art handbook on mobilities, one of the founders of the field of Mobility Studies, John Urry, focuses specifically and explicitly on the tremendous impact of American but also global sprawl on climate change: 'Many new suburbs in the US from the 1980s onwards were built distant from city centres. They were not connected to city centres by mass or public transit. Such *Sprawltowns* depended upon car travel and hence plentiful cheap oil, so newly arriving residents could commute to work and drive about for leisure and social life' (Urry 2014:

588). Suburban infrastructures have become the most visible set of socio-technical assemblages that stand for the ecological and financial crisis of our age. This crisis is looking for a technical, but ultimately a political solution.

The politics of suburban infrastructures refers in the first instance to the way in which McFarlane and Rutherford (2008) have introduced the term but also examine decision-making processes and their consequences, in particular as regards who influences infrastructure-related decisions and how different constituencies are affected positively or negatively by these decisions. What singularizes suburban infrastructures within the range of government responsibilities? Infrastructures are often taken for granted and are even invisible in some cases, which tends to dampen public mobilization around issues related to them. The predominance of the role experts play in decision-making further reduces the impact of the public on infrastructure decision-making, which is often dominated by lobbies with a direct interest in these decisions.

In the larger debate about suburbanization as marginalization, the politics of infrastructure plays a very important part. In fact, in the so-called 'Badlands of the Republic', the French *banlieue*, named such by Mustafa Dikeç (2007), a politics of exclusion has been built, among other things, on the poor infrastructural connections provided by state policy (in addition to and exacerbated by other forms of violence by the police, through racial discrimination or massive unemployment (Dikeç 2007)). The infrastructural disconnect is therefore always in the centre of public policy debates as is currently the case as plans for an extensive network of intra-suburban links proceed under the framework of a Grand Paris (Enright 2016; Nussbaum 2017). In Brazil, similarly, the massive protests of 2013 were very much a response to infrastructure inequities suffered by the peripheralized poor in the country's sprawling urban regions; the social movements of the poor in the peripheries did not just politicize infrastructure, they also 'helped to democratize Brazilian society and to significantly transform the quality of the urban space and of the public services in the peripheries' (Caldeira 2013). In Los Angeles, it has long been known that transit justice/racism was linked to the particular suburbanized structure of the city's communities of colour (Soja 2010). And in Toronto, we have seen huge political issues arise over inequalities of access to infrastructure services in the region's inner suburbs, its so-called 'inbetween cities' (Keil and Young 2014; Young and Keil 2014).

Civil society constituencies mobilize around infrastructure issues to benefit from infrastructures or avoid their adverse side effects. In this context, we have seen the emergence of transit justice movements that challenge institutional and systemic inequities of infrastructure investments and delivery, especially with regards to ethnic or racial but also class and gender groups in cities and suburbs. These movements have become part of a larger movement towards making claims for 'a right to the suburbs', particularly in the old, inner, industrial suburbs that have suffered economic deterioration and/or influx of new immigrant and poverty populations over the last generation (Carpio et al. 2011).

Suburbs also often display an embarrassment of riches when it comes to infrastructure: giant sewer lines, huge security infrastructures in gated communities, multi-lane highways, industrial and commercial structures that are beyond current and sometimes future need. Suburban infrastructures have become symbols of catastrophic and creeping infrastructure planning failure, as it has been the case, for example, in Montreal with its oversized and overly distant Mirabel airport and as it has been the case in shrinking Eastern European satellite suburbs where the lower number of inhabitants have caused atrophy in the area's sewage system, where literally not enough flushing is still going on to operate the system (Bernt 2015).

Stephen Graham has famously spoken about the 'blackboxing' of infrastructures. They are often invisible and unremarkable, even when they are ubiquitous and large. He asks: 'How many of the world's burgeoning billions of urbanites, after all routinely consider the extraordinary assemblages of fuel sources, generating stations, transmission wires and transformers that push electrons through the myriad electrical artefacts of contemporary urban life? Or the mass of servers, satellites, glass fibres, routers' (2011: 67). In many ways suburban infrastructure has been working as the ultimate black box. As a 'zone' with a 'disposition' (Easterling 2014) that is partly self-sufficient and has a purpose onto itself and is sometimes there to serve other parts of the urban region (including the downtown and other suburbs), the suburban infrastructural networks and grids are massively unknown and in need of better understanding and explanation. Suburban infrastructures suffer from the same simplification and stereotyping, the same neglect as the suburbs themselves.

RETROFITTING SUBURBAN INFRASTRUCTURE FOR/IN POST-SUBURBIA

In the remaking of the suburban landscape which is on the agenda of today's urbanism, it will have to be seen what role remodelled and repurposed suburban infrastructures can play in helping this process along. Ellen Dunham-Jones, a thought leader in suburban renewal, has pointed to the 'terrific infrastructure and a relatively central location in a now-expanded metro area' that might be found in ageing suburbs (Dunham-Jones 2015). If we go one step further and look at housing as a suburban infrastructure (Young, Wood, Keil 2011), we might include the renewal and energy retrofitting of suburban high-rise towers in the fold (Hume 2015). Even more advanced, we might think of the entire suburban landscape as a Large Technical System (LTS) in need of energy renewal (Dodson 2007; 2014; Dodson and Sipe 2009).

Cities and suburbs worldwide face more and more diverse infrastructure problems than ever; at the same time, solutions to protracted issues related to how people get around, access energy and water and communicate are more numerous and more interconnected than ever. Technological innovation and governance sometimes coincide and sometimes cause major disruption to existing business models and the organization of work. Some of the companies that drive the process of innovation in the mobility field, for example, are major software and IT firms such as Facebook, Apple or Google; others, such as Uber, Otto or Lyft, are startup giants that emerged quickly over the past decade to claim a dominant spot in the market and in the public debate. The corporate consulting firm McKinsey summarized the range of expected innovation in urban mobility in a comprehensive report in 2015 to analyse the incipient 'urban-mobility revolution' they see occurring. They identify seven factors that will influence sustainable mobility in cities and suburbs in the near future. Among those that directly affect suburban infrastructures they name privately owned vehicles for which they identified 'four major technological trends [they see] converging: in-vehicle connectivity, electrification, car sharing and autonomous driving' (Bouton et al. 2015; see also Guerra 2016). The report predicts that in both established and emerging urban regions – and we assume that much of that affects the vast suburbanizing terrain in today's metropolitan regions – those four trends will interact variably with existing forms of governance and regulation to create a host of new infrastructural constellations that will change the way suburban land use and mobility interact. In this process, the classical (American) suburbanization based on automobile dispersion and zoning is called into question as a sustainable way of building cities (Bouton et al. 2015). Instead, the report celebrates emerging new models such as that of transit oriented development (TOD) in 'Chengdu in southwest China, for example, [where] a new satellite city is being built for 80,000 people that could serve as a model for a modern suburb. Instead of a layout that makes it necessary to drive, the streets of what will become Tianfu District Great City are designed so any location can be reached in 15 minutes on foot. Motorized vehicles will be allowed on only half the roads; the rest are for walkers and cyclists' (Bouton et al. 2015). It may be necessary to add here that, in contrast to the typical American subdivisions, many European suburbs that were built inbetween existing village and town structures have always had this kind of multi-modal connectivity and have served as models for suburban design in China, such as the automobile-industry new town of Anting near Shanghai, which was built in the form and density of a German suburban town.

Infrastructures in emerging sub/urban environments are potentially innovative beyond their place of origin. This is true, for example, for the tremendous success of the TransMilenio Bus Rapid Transit project in Bogota, Colombia (Montero 2016), of the much admired system in Curitiba in Brazil, of the regulation of private automobile use in Singapore and the subway system in Shanghai (Shen 2015). Those and other examples have now ascended to be the leaders of the global debate on how to move the urban periphery and have challenged the colonial narrative that used to imply that innovation came from the North with its headstart in engineering and technical knowledge and application. This narrative has been changing lately as sub/urbanization patterns have shifted more generally, with stronger growth and more rapid development occurring in the Global South. Ananya Roy has also pointed out that there is now much south-to-south inter-referencing in these learning processes: 'While such practices can be read in the register of urbanism, it is important to make note of their worlding character. (…) inter-referencing…reveals how the production of urban space takes place through reference to models of urbanism. In India, urban planners and city elites turn to Asian models of urbanism such as Shanghai and Singapore for reference and inspiration. Such forms of inter-referencing make possible the transformation of urban disorder, the dystopia of the Global South, into civic order and postcolonial pride' (Roy 2011: 10). While there are many obvious successes of the inter-referenced technological awakening of an infrastructure-based urbanism, critical voices have also noted that such measures might in fact increase existing inequalities in emerging cities of the Global South. In Rio de Janeiro, for example, a train, a highway and a bicycle speedway connect the city centre with the luxury gated communities of Barra de Tijuca in the South, while little has been done to connect the poor northern suburbs. Introducing cable cars to connect favelas and barrios on the periphery certainly adds to the mobility choices there, but such investments have also been criticized for spending public money on a showpiece project while leaving other issues unattended to. The cable cars may need to be understood at least in part as 'beautifying the slum' (Rivadulla and Bocarejo 2014). There are certainly many examples from the Global South that come to mind in this context, among them the building of the Delhi Metro, which has been as much about building an image as it has been about building an actually people-moving subway; it also bypasses poorer areas of the city and cements existing inequities (Siemiatycki 2006).

In this context, we encounter another major theoretical contribution in terms of infrastructure theories from the Global South: the notion that people themselves *are* the infrastructure (Simone 2004). This latter provocation leads the debate away from the Western technological and material preoccupation when it comes to the infrastructural organization of cities and suburbs. We will need to understand the ways in which the focus of the entire debate on providing infrastructures (water, mobility, etc.) has been unduly on the fixed capital of the core as opposed to the soft infrastructural capacities and innovations located in the periphery (Swyngedouw 2004). This is, of course, a major question of epistemological optics as we have learned. The 'infrastructural ideal' was never achieved in most sub/urban places around the world but has remained a powerful fiction in the way in which the peripheries of cities continue to be provisioned (Kooy and Bakker 2008; Bakker 2010).

Suburban infrastructural innovation is not just a factor in emerging metropolises of the Global South and East but also in established urban areas. Commenting on Paris's planned suburban subway extensions, Paul Konvitz notes: 'The rate of return, through higher per capita income, spending and higher corporate profits, makes this a wise investment with a one-hundred-year horizon. Cash-rich, the private sector can act fast when government with a vision sets the strategic framework for the urban future' (quoted in Florida 2016). The social geography of transit innovation and disruption is peculiar and contradictory. It is that way as it calls into question the status quo in the 'transport land use feedback cycle' (Bertolini 2012) as it has existed since the beginning of mass suburbanization when suburban automobile and transit infrastructures were created to deal with what amounted to mostly a one-dimensional and uni-directional commuter stream from the suburbs to the central city and back again. Today's metropolitan transportation demands are different, commuter patterns change and employment rhythms have lost their predictability. In Paris, for example, 'the westward shift over the past two decades of the center of gravity of employment away from Paris and toward the suburbs presents a problem for static infrastructures and the less mobile residential markets. Deep divisions exist between home and work for many inhabitants. These divides are only set to grow with commutes in the near suburbs predicted to rise by 15 percent by 2020' (Enright 2016: 108–9).

Under these conditions, suburban infrastructures can only be understood in a complex pattern of contradictory dynamics that include issues of housing and employment, warehousing, commercial and other geographies. This is perhaps nowhere as visible as in the current crisis around the expanding tech industry in the San Francisco Bay area where growing employment, changes in consumer preferences, corporate locational decisions and a severe housing crunch caused by restrictive policies in the city and the suburbs add up to a major crisis (Cutler 2014). While the so-called Millennials are considered mostly a generation that prefer urban over suburban dwelling and have contributed significantly to lower car sales in established cities, many of them work for tech companies that continue to colonize suburban areas such as Silicon Valley where Facebook, for example, has been offering employees thousands of dollars if they locate close to their headquarters in the periphery. Alternatively, many companies in the vast suburban expanse of Silicon Valley have private bus services to allow workers to commute to other neighbourhoods in the Bay Area, especially San Francisco. Consequently, communities everywhere along the newly established routes between the tech workplaces and the housing districts in gentrifying neighbourhoods have organized to stem what they believe to be wave of change to mobility and land use that will disadvantage existing populations and workers in traditional labour relationships; workers in the new bus companies have unionized to strengthen their bargaining position vis-à-vis their corporate employers (*Guardian* 2015). Many of the new technologies and their social networks are associated with the push towards re-urbanization (Cutler 2014).

CONCLUSION

The process of explosion in the Lefebvrian sense can be seen concretely as techno-social developments such as the ones discussed in this chapter: 'the increase in mobility, the de-centralization of residential and commercial functions, the explosion of digital communication, the fragmentation of families and firms' (Balducci 2012: 4). These real and functionally separated processes 'are now displaying their combined effects and producing a profound reconfiguration of 'the urban' and its structure' (Balducci 2012: 4). Balducci also insists that 'the new configuration of urban spaces is based upon co-existence rather than upon substitution or opposition' (2012: 5) and that the development of and in the metropolitan periphery 'is not a development *against* the wealth of central cities, rather it is part of the same new form of urban development. Peripheral centers grow due to their internal vitality and attractiveness and to the delocalization of activities from the central cities, thus creating a new spatial pattern' (Balducci 2012: 5).

Tower Density in Tallinn, Estonia

8 | The Urban Political Ecology of Suburbanization

'These spaces are produced. The "raw material" from which they are produced is nature. They are products of an activity which involves economic and technical realism but which extends well beyond them, for these are also political products and strategic spaces.' (Henri Lefebvre 1991)

'Sundered apart, nature and society die in reciprocal conceptual torpor.'
(Neil Smith 2006)

The epigraphs that open this chapter name the inevitable imbrication of the urban and the natural in processes involving socio-natures and urban metabolisms (Swyngedouw 1996). A growing literature on Urban Political Ecology (UPE) has examined these relationships and has been 'not so much concerned with the question of nature *in* the city, but rather with the urbanization *of* nature, i.e. the process through which all types of nature are socially mobilized, economically incorporated (commodified) and physically metabolized/transformed in order to support the urbanization process' (Swyngedouw and Kaika 2014: 462–3). UPE explores existing power relations and views cities as historical products of human-nature interaction. One of the main themes underlying most of the UPE literature, as Neil Smith suggests above, is that the urban and the natural are not seen as separate entities, but rather as intertwined and inseparable from one another (Heynen 2014; Keil 2003). This is in direct contrast to historical ideas about the city and nature, especially the clear separation between town and countryside in nineteenth-century capitalism (Wachsmuth 2012: 508).

In this chapter, I discuss *sub*urban political ecologies. Suburbs as places, suburbanization as a process and suburbanisms as ways of life have variably been seen at the centre of today's debates on climate change, sustainability and even resilience. Suburbs as places are seen as unsustainable, suburbanization is viewed as a process that has to be arrested and suburban ways of life are seen as incompatible with the challenges of a heating planet. In this chapter, we approach these salient topics through three connected lenses: the question of density, the notion of boundaries and the challenges associated with the Anthropocene. In all three cases we will see that the suburban is a complicated terrain on which new socio-natures are negotiated. Urban peripheries and the processes that make and sustain them are not simply culprits in climate change. They need to be viewed as complex societal relationships with nature from which new dispositions for action may emerge.

Traditional UPE has only been partially useful for this endeavour. Just a few participants in the UPE debate have focused on suburban issues of all kinds (e. g. Jonas et al. 2013 and Pincetl et al. 2011; Robbins 2007). My interest here is what contribution UPE can make to the theorization of the aggressive suburbanization processes that define today's urban growth and that have been the subject of this book. To begin, UPE conceptualizes the city as a sub-system located within a larger socio-spatial system (an urban region) and the notion of metabolism highlights the interaction between and among multi-scalar systems (Heynen *et al.* 2006). Particular place-based metabolic relationships are embedded in and constituted by external scalar and topological relationships that define the metabolisms in a particular region. For example, immigrants to new suburbs, capital from abroad used to finance the building of subdivisions and commodities provided through the world market define the boundaries and perforations of metabolic relationships at the suburban frontier. But those metabolisms have to be stabilized and reproduced through local action under regional regulation. Densities are a central terrain for negotiation of what is urban and what is not. What is city and non-city is defined somewhat through the political and social ecology of boundary-setting exercises such as greenbelts. And our understanding of survival in an Anthropocenic world is largely connected to the ways we live in an extended form of urbanization. While we can imagine the metabolism of the urban region as a seamless web of human and non-human life with metabolic streams connecting organisms, we also must recognize the obstacles and facilitating channels by which those streams are conditioned. Let us first look at the density debate.

TOWERS IN THE PARK, BUNGALOWS IN THE GARDEN: PERIPHERAL DENSITIES, METROPOLITAN SCALES

In terms of sub/urban political ecology, we have recently become accustomed to the idea that 'sustainable urban development' is better achieved in compact dense cities than in sprawling low-density suburban forms of settlement (Keil and Whitehead 2012). Yet we may have to rethink, again, such simple opposition of densities as a single marker of sustainable metabolic relationships. As has been argued in this book, suburbs have traditionally come in all sizes and densities. Nonetheless, they have been mostly imagined as places of low density, privatism and automobility. Such singularity of focus on a particular form of suburban life has been myopic and has limited the view towards suburbia as a global phenomenon. In as far as the low density suburban morphology has been seen as unsustainable, the suburban project overall has (often correctly) also been marked as a major cause of disruption in the socio-natures of our time. But the density story may be more complicated than that.

In this section, then, I will disentangle the real-existing suburbia from some of its enduring myths related to densities. I will argue that while densities in suburbs have always been mixed and diverse, the emphasis on a particular form of US-style individualized suburbanization has led to a focus on a cultural politics of social conservatism, property ownership and privilege which has obscured alternative social and built forms in suburbia. It has also obscured some of the pathways through which post-suburban metropolitanism has to find a solution to some of the most pressing social, economic and environmental issues we confront in urban regions today. On the reverse side of these representations, the appearance of the stereotypical suburban *morphology* as an ideal(ized) *community* has contributed to hiding poverty, diversity and change and to clouding the view towards extant differentiations inside suburbs (Touati-Morel 2015). This partial blindness complements the normative position taken by many planners and urbanists that see the density of suburbia as a problem. This has led to the hyperdensity proposition: recent pleas for models of urbanism that involve an even higher concentration of humanity in dense urban cores at a density of 30 units per acre (Chakrabarti 2013; Glaeser 2011). In this context, I am disregarding the logical opposite of the hyperdensity proposition, the normative preference for low-density settlements. This preference, mostly identified with a certain strand of American libertarianism has little intellectual and no political merit or plausibility (Cox 2012; Kotkin 2013).

In the historical settlement of the (soon to be suburbanized) countryside, based on the principle of enclosure to a considerable extent, a new narrative was woven with the threads of a 'new symbolic economy that had a direct influence on the image and imagination of national territories' (Sevilla-Buitrago 2014: 251). To this imaginary, the single-family home was central. That it also fit nicely with the growing demands of a mass-produced form of housing and the desires of the development industry to maximize their profits on a grid of tract housing cemented the role of the bungalow-mansion spectrum of houses at the core of the suburban model (Ford 1994). This ideal ultimately also infused the massive suburbanization that characterized America in the twentieth century. And, as Keller Easterling has noted '[a] design idea for suburbia becomes more powerful when it is positioned as a multiplier that affects a population of houses' (2014: 84). The house became the relay station of the Fordist-Keynesian regime of accumulation, a position that lingered into the current age of Metroburbia when consumption of space and gadgetry knows no bounds (Knox, 2008). At the same time, the libertarian ideal of the individual house as the centre of societal life ran up against the strong urge by modernist planning and state bureaucratic institutions to order and regulate the urban experience (De Meyer et al. 1999: 17).

Suburbanization has been seen occurring in line with the spread of a 'postindustrial society' (ibid. 24), 'the embodiment of a new kind of city' (ibid. 27). This new city – dispersed, multi-focus, discontinuous, variably dense and multicentred city, not oriented towards the traditional core – has been called 'post-suburban'. Post-suburban changes have now been registered across the globe, albeit in various forms and to varying degrees. In some regions of the world, like in the states of the former Soviet Union, once dense peripheries are now developing patterns more reminiscent of the West. Accordingly, single-family home construction in the peripheries of the region exploded after 1990 and so has the presence of big box commercial developments typical of the in-between cities of the West (Stanilov 2007: 179–86). This is resulting in a mixed landscape likened metaphorically by Hirt and Kovachev (2015: 193) to a 'multi-headed deity' in the tradition of Eastern European pagan religions 'looking to past and to future and to the four corners of the world'. In his work on the post-suburbanization of the Global South, Terry McGee has concluded: 'that suburbs are best defined as a category of settlement that is one of the many types of the built environment of housing settlement types, commercial and industrial spaces, as well as infrastructures that include high-rise apartments, town houses, condominiums, family homes and illegal settlements that are now part of the emerging fabric of an urbanized world' (McGee 2013: 25). Similar observations can be added from the experience in India (Gururani and Kose 2015) and Africa (Bloch 2015; Mabin 2013). In China, a newly sub/urbanizing country, where Wu (2013; 194) sees a process of 'urbanization of suburbs' under way, 'suburban development is more about constellations of urbanization and suburbanization and it is heterogeneous in terms of population composition'. In her case study of Wangjing in suburban Beijing, Fleischer (2010) has registered a hierarchy and typology of different morphologies that have rapidly congealed into a mature and diverse suburban landscape over the past 20 years.

Fran Tonkiss (2013: 43) has noted that in a global perspective 'debates over density and sprawl become not only sterile or semantic, but also increasingly irrelevant, for the majority urban experience'. However, density, or more specifically higher rather than lower density, is a concept that enjoys much love in urbanist circles. It is often seen as a necessary precondition for sustainability and even resilience. It has become a rarely questioned normative objective. Densification and intensification – often paired with the notion of compactness – are desired processes of urban planning and development. They are the mantra of central city-based urban environmental policies ostensibly geared towards a twenty-first-century urbanity tied to pedestrianization, cycling and transit as mobility options of choice. A battle for 'sprawl repair' and for the urbanization of the suburbs has produced a cottage industry of travelling urbanists, pundits and planning consultants that range from the serious (Dunham-Jones and Williamson 2011; Jessen and Roost 2015b) to the naïve (Gallagher 2013). The thrust of such enthusiasm for density and compactness is easy to understand if not somewhat contradictory as it often neglects the complex problems related to a single focus on an (abstract) concept of density at the expense of other considerations related to sustainability and resilience (Charmes and Keil 2015).

The preference for density and compactness, so strong today, clearly hasn't always been there. In fact, the modernists had proposed to disassemble the city and to reassemble it in new rational landscapes of metropolitanism that may have had density cores but would be rather dispersed overall (De Meyer et al. 1999). Generally, the suburbanizing region put in motion a trend by which 'the areal extent of the built-up city has grown faster than population' (Bunting et al. 2002: 2534). At least since the 1990s the functions of urbanity and density have become disentangled (De Meyer et al. 1999: 27). Mixing of functional elements and densities becomes the norm in the post-suburban inbetween city across an urban region punctured by edge cities and technoburbs (Garreau 1991). In this process, in Canada, for example, 'wasted density' (i. e. islands of high-density residential areas stranded in a sea of low-density single-family homes) is seen as an inevitable outcome of metropolitan spatial distribution (Filion et al. 2006). As this new multi-density, multi-function urban region has emerged over the last generation, it has consistently defied systematic analytical, cartographic representation and planning intervention (De Meyer et al. 1999: 30–8).

Density deserves another look in order to understand some of its inbuilt contradictions (McFarlane 2016; Tonkiss 2013). We might ask questions such as: is there 'good density' or 'bad density'? Is the term as simplistically positive as is often suggested? Are processes of intensification value free apart from their positive effect on the environment? I offer a few observations below that may have to be taken into account when discussing the term in policy and planning debates. My main interest here lies in illuminating the notion of density in the conversation about urbanizing, i. e. densifying the suburbs.

If we talk about suburbs and density, we need to first recognize that both terms are 'chaotic'. They defy clear definition but everyone uses them with a certainty that often leads to heated debate and misunderstanding. There are many versions of both suburbs and density. In the growing debate on what to do with suburbs and suburbanization, the normative preference of most self-declared urbanists is clear. Suburbs are deemed unsustainable, costly and unjust due to their sprawling use of land, their increased infrastructure cost and segregative socio-spatial structures. The cultural prejudice against low and medium density also has a certain colonialist tinge. Planning professionals and urban visionaries, often themselves at home in gentrifying neighbourhoods of city cores, seem to know best what's good for the suburbs and the planet. They also tend to disregard cultural preferences for less dense living in contexts other than North America or Europe. South to South inter-referencing in city building (both good and bad) is often overlooked in the process. In fact, most North Americans and Europeans would abhor the already existing low-rise densities to be found in African squatter settlements and slums. Lagos, Nigeria, for example, the fourth densest large city in the world by some measure, has a density of 18,150 people per square kilometre but hardly the built form and networked infrastructure to support the kind of urban life North Americans or Europeans would associate with Good Density (CityMayors Statistics 2007). In Maputo, often touted as one of the success stories in African urbanization, still eighty per cent of the population live in slums (Campbell 2005). The discourse there is not density but poverty, not abundance but need. Tonkiss (2013: 38) admits accordingly: 'The problem of density sits in an uncertain place in-between: the point at which density becomes overcrowding, especially in slum settings, is never quite clear.'

Density as a site-specific quality is almost meaningless if one does not look at the broader societal context and patterns of use as well. A person can live in a compact, dense, walkable neighbourhood but work a long drive or transit ride away from it. In an age when families are not traditional and don't follow the breadwinner/homemaker model of the (sub/urban) past, choice of residence is no longer a question of linear live-work relationships. If there are double income earners in one family, they likely work at different locations in the urban region. The location of their home will have to be a compromise. Many educational institutions of higher learning are now in metropolitan peripheries (Addie, Keil and Olds 2014).

Another context is wealth and class. Rich people often live in dense neighbourhoods with lots of cultural, culinary and commercial amenities. But those places may not be accessible to public use, they may be gated or concierged, protected from public access, or priced out of reach. They also may not be the most ecologically sustainable by themselves (Cohen 2014: 151). Density can be one of the best ways to concentrate wealth and power; and density can be one of the best ways to concentrate poverty and oppression. Wealthier people may be living in dense quarters but they are also more likely to travel, to own a cottage, to drive multiple and bigger cars. In that case, residential density may be good for their home community but it makes no difference to the planet. Poorer people may be forced to the periphery by gentrification; they may have multiple part-time jobs that force them to be mobile across a sprawling urban region. They also have less political clout and their influence on the overall urbanist choices in their city tend to be minimal by comparison with the often professionally trained, wealthier inhabitants of the dense, bourgeois neighbourhoods downtown. If they live in tower neighbourhoods, their home is vilified by a discourse that problematizes that particular form of living. If they live in the bungalow city, their sub/urbanist aspirations and automobilist lifestyles are considered part of the problem. The current discourse on the dense city marginalizes both experiences and their inhabitants. A dialectics of density and sprawl is the reality of the regional city while the discourse on density pits one against the other in a rather unproductive manner.

Surely, sixty-one per cent of all Americans now live in single unit detached housing on about an acre of land as Chakrabarti (2013: 26) has shown. That is a real problem in so much as in North America, in particular, those living arrangements are supported by the most unsustainable economy in the world, in terms of energy consumption per capita, waste and so forth. The same kind of densities may not be quite as disastrous in an economy that is set up in more modest terms or that have full-fledged ecological modernization processes going on that affect energy use, mobility, etc. in other ways (i. e. many Swiss and South Africans live in single-family structures in low to medium density but in both – rather different – countries, much less energy use is needed to live in those structures and to move people around). The density-centrality equation abstracts from a lot of other factors in the urban region. The complex geographies of power and wealth have to be factored into the equation instead of focusing merely on one side of the problem: the sprawling suburbs.

Given such preference for a particular kind of density, a few critical observations are in order. Planners and urbanists coalesce in a double movement that (a) condemns existing and future suburbanization and (b) proposes more compact re-urbanization initiatives in return. In this double movement both steps are problematic. In the first instance, assumptions are made about the suburbs and suburbanization that may not hold up to full scrutiny. The proposed remedy for the damages done by suburbanization contains an amalgam of mostly site-specific and urban form-oriented measures that leave much context unconsidered. They operate on the scale of the house, the street and perhaps the neighbourhood and leave broader scales and topological relationships of those units unconsidered. The region and the world remain the blind spots of urban density. The normative debate on urban density is relatively uninformed of the complex 'suburban political ecologies' specifically and the larger city/countryside dialectics generally that sustain it in a world of fully extended urbanization (Ajl 2014; Angelo and Wachsmuth 2015).

Fed by normative positions on density like the ones just discussed, there are three assumptions that commonly underlie the critique of the sprawling suburbs.

(1) The 'undercomplexity' assumption. This is based on a very uniform picture of what suburbanization consists of today.
(2) The 'suburbs don't change' assumption. While the inner cities are depicted as lively laboratories of urban change and variable densities, the suburbs are caricaturized as unchanging and static.
(3) The 'suburbanites-are-passive' assumption. Like the suburbs themselves, their residents are often considered passive, self-interested, private and even worse opposed to the density push that comes from the inner city.

Contravening these normative assumptions, Eric Charmes and I (2015) have critiqued the density prerogative from three angles:

(1) the politics of densification. Densification is not neutral but it is contested across various political camps. The environmental arguments in favour of densification are often plastic and dependent on the speaker's interest and subject position. The argument that denser cities are more sustainable than their lower density counterparts only holds if many higher scale issues of energy use etc. are ignored. Smart growth and new urbanism have now become *de rigueur* in development discourse as land developers, builders and local political elites sell growth (and their ambitious plans) with a green strategy.
(2) Moreover, in the density debate morphology matters. Density per se does not say much about urban form. There is a wide variety of ways by which to achieve similar types of density.
(3) The existing suburban densification regimes are not all the same. They differ substantially from each other. Both physical landscape and socio-political governance present a wide spectrum. Class and other social status differences are important in this context.

In conclusion to this section we can maintain that real existing density is highly uneven and suburbs are as diverse as many inner cities. The processes currently set up to increase density carry with them a high potential for reinforcing existing inequalities and positive benefits for sustainability are questionable. Stepping outside of the dense/city – sprawling/suburb mode of thinking and filling the gaps between the high-rise condominiums downtown and the single-family homes in the periphery remain crucial imperatives. In a scenario where millions of newcomers are yet to arrive in many already densely populated urban regions over the next generation, we cannot afford to look at either suburbs or density in the ways we have to date. It is time to change our view of the suburbs without confronting the suburbanites with rigid normative statements on density. Both in substance and process, the densification of the larger metropolitan region must be a democratic and differentiated affair that defies the stereotypes used in much public debate and planning discourse. The proposed departure from a dichotomic view of urban density versus suburban sprawl reveals a

Greater Golden Horseshoe Greenbelt, Ontario

much more intricate tapestry of morphology in the urban periphery. The iconic images remain central to this re-politicized margin of the urban region: the tower and the bungalow. Yet, they also are beginning to shift in both material reality and meaning. Ultimately, it is important to recognize that the towers were in the public park and the bungalows were in the private garden (Poppe and Young 2015). This traditionally points to a register of potential conflicts around the scaling of very different political projects, one public, one private. Today both are placed next to each other at an even larger scale of suburban political ecologies: the post-suburban landscape.

BOUNDARIES OF SUBURBAN POLITICAL ECOLOGIES

'The urban looms on the horizon as form and light (an illuminating virtuality) as an ongoing practice and as the source and foundation of another nature or a nature that is different from the initial nature.' (Lefebvre 2003: 108)

In this landscape, a reordering of the classical city-nature relationships that are subject to examination in UPE is taking place. This includes the setting of practical boundaries between the social and the natural and it occurs normally, as Lefebvre surmises above, in the light cone of urbanization. Suburbanization itself is an exercise in boundary setting.[1] It is both a statement against the city from where it distances itself by providing new housing, employment, entertainment, infrastructure, etc. beyond the traditional core and a statement against the rural beyond. Suburbanization remains the common boundary of the metropolitan region with its natural outside, whether that is in the form of 'feral' nature as it is in the Canadian north without much of a boundary between the vinyl suburban world of universal subdivisions and the bogs and swamps of an unforgiving nature just beyond (Shields 2012) or the desert of the American southwest, or whether the outside is a farmed agricultural landscape or industrial region. In this section, I explore the suburban political ecologies of boundaries and specifically greenbelts as conceptual boundary-setting exercises that engage with the suburbanization process.

Suburbanization redefines urban boundaries conceptually and physically. Chief among those is the constantly changing relationship of nature and urban society. Metropolitan governance works those boundaries through buffer zones, wedges, fingers and urban national parks that contain and govern suburban expansion and/or conserve natural or agricultural landscapes around metropolitan centres. A global movement, the greenbelt initiatives are now imposing themselves on the political discourses of a variety of urban areas around the world. The transition from city to countryside has always been a critical interface of human existence and a defining marker of urban political ecologies. Since antiquity, the inside/outside problematic has been a central concern of urbanism: walls, moats, gates. Boundaries have forever defined the urban. Historically, the clear break between the city and the hinterland has come in one of two ways: the *edge* (as in the German use of *Stadtkante*) and the *sprawl*. Clearly, the former has historical significance as a defensive built environment. A sprawling mesh of settlement can hardly keep the enemy at bay in a warfare based on brute physical conquest! But the boundary between the urban and the suburban has also always been set by social process and socioecological metabolism.

Suburbs have historically been lines of definition between the urban and the rural (Teaford 2011: 15). Boundaries, then, are historico-spatial products of relationships between urban activities and non-urban activities. Needless to say this kind of dichotomous thinking needs to be interrogated in a post-suburban world (Phelps and Wu 2011). In post-suburban environments the clear inside-outside relationships of the concentric suburbanization process is dissolved into various competing and mutually defining dynamics of producing suburban space. Old and new, rich and poor, people of various cultural backgrounds are segmented into a polycentric fabric of multiple dependencies. Suburbanization is now so pervasive that it is present on both sides of the greenbelts as leapfrogging of development beyond the designated conservation spaces is a wide spread occurrence.

Following Walter Prigge's (1991) reading of Walter Benjamin greenbelts are 'thresholds' more than simple boundaries. A threshold has a *spatial* quality, it is not just a line but a space for multi-facetted negotiation. Crossing the threshold changes the perspective. The greenbelt as threshold points to the central contradictions of modern metropolitanity of which suburbanization is a major factor. Relatedly, Laura Barraclough has examined a 'rural urbanism' in Los Angeles's San Fernando which she says has a 'complex past in which rurality, suburbia and urbanity have coexisted, often tensely...the Valley's distinctive mélange of rural, suburban and urban landscapes attests to competing visions of the kind of place the San Fernando Valley has been and will become – visions that are intimately linked to the ways in which racial, class and national identities are being negotiated in Los Angeles and the Metropolitan U. S. West' (Barraclough 2011: 3).

The developmental dynamics of the boundary setting that greenbelts perform in an era of global suburbanization are vastly different in various places (Amati and Taylor 2010; Carter-Whitney 2010). In the old industrial landscapes of Birmingham, Manchester and the Emscher (Basten 2016), grass grows literally over the ruins of an obsolete spatial fix. In the booming regions of Toronto (Keil and Macdonald 2016), London (Gordon 2016), Frankfurt (Keil and Ronneberger 1994), Delhi/Gurgaon (Gururani forthoming) and Seoul (Hae forthcoming), greenbelts are expansion spaces and projection screens where the global city is 'going up the country' (Keil and Ronneberger 1994).

The landscape beyond the city in which the greenbelts have been established has long been a rural landscape. The infrastructures that have been laid out there have facilitated agricultural production and, in the cases of the blue and green landscapes of Belo Horizonte, in the Black Country around Birmingham and the Emscher valley in the Ruhr, also industrial or mining activities. In this sense, the greenbelts' matrix has been inscribed in physical and social terms as non-urban. Yet today's greenbelts have to be seen not as qualitatively different from urban structures and relationships but rather as conjugations of post-suburban relationships and terrains of novel suburban political ecologies. The design and use of those greenbelts today are taking shape not against, but in conversation with, the push of suburban expansion that surrounds those greenspaces. The use of infrastructures in the greenbelt is not anymore governed by the primacy of agricultural pursuits but by the interests of urbanites in real and imagined green landscapes. 'Para-agricultural' uses (horse farms for riding by urban kids for example) trump traditional agricultural pastimes. The musealization of industrial and agricultural histories replaces the actual industrial or agricultural production in greenbelts. In this sense, the suburban boundary is geographic, historical and ecological.

This relates ultimately to the relationship of agriculture and the city: the infrastructure of the agricultural mode of production now becomes a new spatial fix for the post(sub)urban realities of the regional metropolis. What effectively lies beyond the suburbs and is articulated with their spatial forms and social processes is in most cases a concrete, farmed landscape, not 'nature' in any abstract sense. The greenbelt holds as a boundary for suburban development and becomes the dreamspace-threshold for alternative uses of green space. The firm boundary does not exist, the suburbs have arrived in 'nature'.

The *sub*urbanization of nature corresponds to an imbrication of the suburban with nature along perforated boundaries. The territorial boundary between greenbelt and suburbia separates and negotiates an increasingly complex set of productions of socio-natures that belie the original simplicity of the city-country dichotomies in the earliest imaginaries associated with greenbelts in the early twentieth century in the

UK and in Germany. The interface of the suburban and the natural environment beyond is an important frontier of global suburbanization, a defining feature of today's extensive urbanization and part of the governance of suburbanization. The specific metabolic and discursive constellations engendered by massive suburbanization and the accompanying attempts at limiting these expansions through environmentally themed and legitimized spatial boundary-setting exercises provide a prime lens through which to understand processes of post-sub/urbanization more generally.

THE URBAN POLITICAL ECOLOGY OF GREENBELTS

The urbanized region has begun to develop as a patchwork. As Tom Sieverts (2003; 2011; 2015) has reminded us, even the green spaces beyond the traditional city cores are now perforated in dramatic ways and have become part of the inbetween cities that are typical for this age. The metaphor of the belt is one of control over flow. Can this enclosure really be a boundary that holds in a world of perforated borders? The greenbelt, as a negotiated space of societal relationships with nature that connect urban and non-urban activities, becomes a canvas for the production of (social/urban) nature where suburbanization and urbanization are enclosed, enabled and energized. Metabolism is ultimately about the localization of flows in specific socio-natural processes in situ (we can think of this in corporeal terms or in terms of technological systems). This may cast the process of sub/urbanization in terms of how socio-natural flows accommodate the built and social environment, undoubtedly leading to the exploitation of natural ecosystems while, in turn, nature is (discursively) preserved and (legally) protected. Greenbelts are an important planning mechanism as they not only control growth and flows, but they are also considered to be important landscapes for urban resilience (Buxton 2011). In this context, greenbelts have been used not only as a planning tool to control urban growth and prevent sprawl but also to create green space for sub/urban populations. They create a (real and imagined) rural and urban threshold.

THINKING ABOUT POST-SUBURBANIZATION

Boundaries in this context, as historic-spatial products of relationships between urban activities and non-urban activities, need to be rethought in a post-suburban world. In post-suburbia, the rural and the urban are sublated into a new landscape of which city, boundary and fringe are mere elements. Today's greenbelts are not qualitatively different from the city but congealed conjugations of post-suburban relationships, temporary sedimentations of the confrontation of suburbanization with its outside. They take shape *in conversation with* the push of the surrounding suburbs. The use of infrastructure in the greenbelt is no longer subject primarily to the primacy of agricultural pursuits but by the interests of public and private actors who are based in the city in real and imagined green landscapes. Greenbelts today are brought into play and contested through discourses of value and valorization of nature. They are viewed frequently in terms of efficiency and economic value as is common in neoliberal societies (Macdonald and Lynch forthcoming). At the same time, the moral and political geography of the greenbelt is potentially complex. It is as of yet unclear whether and to what degree such a 'greenbelt of the twenty-first century' can stem the tide of suburbanization – or at least qualify and define it – given all the pressures and demands placed on it.

The greenbelt becomes part of a canvas of new metabolic relationships (or if you wish, ecosystem services) in sectors such as mobility, water, food, recreation and even culture. This makes the greenbelt a constitutive piece and even an enabler in the post-suburbanization of the region as new and ageing suburban landscapes evolve into a more appropriate and perhaps resilient rapport with their assumed counterpart in the regional landscape. This forges as of yet underdeveloped societal relationships with nature that connect urban and non/urban activities in novel ways. The conventional 'consumption' of land through development is transformed into the 'production' of a new social and urban nature where post-suburbanization processes are enclosed, enabled and always emerging. Rohan Quinby (2011: 128) has warned that 'the spatiotemporal ecology of the postmetropolis has subsumed the entire built environment of North America, displacing the logic of the traditional city within a suburban-like order of horizontality and dispersal'. This has come with a simultaneous 'explosion of romantic and ideological writing on cities that does little to uncover the connection between our environments and capitalist power' (Quinby 2011: 129). The currently widely practiced wishful thinking of a renewed urbanism in the centre has done little to theorize the wider metabolic relationships that are at the basis of the reproduction of urban life in the inner city, its inbetween regions and its outer suburbs. We can now conclude that, understood in the manner suggested above, the greenbelt is not a static enclosure, a timeless emerald band in an otherwise alienated and destructive neoliberal cityscape of horizontalized high density and commercialized space (Quinby 2011). Much more, it is part and parcel of a performative post-suburban renewal that has the potential to point beyond the drab, dichotomous status quo of destructive hypergrowth in the city and green musealization outside of its conventional borders. Recognizing the important regional and global metabolisms that sustain the urban political ecologies of the greenbelt and its surroundings opens the conceptual door to a new way of thinking about post-suburbia as a terrain of liberation (Keil and Macdonald 2016).

GREY TRANSITIONS

The dynamics of the suburbanization process itself, with its now distinctive patterns of diversity, density and development inscribe themselves onto the nature beyond. Not uniform in demography and economy, the periphery offers a variety of access points and connectors with what lies outside. The threshold between suburbanization and the greenbelt, therefore, is a zone for a grey transition. For this I take inspiration from Oren Yiftachel's (2009) term 'grey space' which he considers distinctive of the political geography of urban informalities he sees as characteristic of today's urbanized world. The spaces that concern me for the present purpose share the shadowy and 'outside of the gaze' qualities of those informal constellations. But they are also among some of the most surveilled and monitored, as well as regulated areas of our urban environment. They are domains of hard speculation and soft desires of the entire urban region in one grey zone, one sphere of projection: 'horsification', post-Fordist economic expansion, the endzone of the suburban dream, the co-presence of privilege (the old super-rich and the new super-rich) with a minimum wage labour force in the suburban warehouses and factories. Grey spaces are contested territory. For the excluded and powerless, they are spaces of unregulated opportunity but also refuges created by necessity. The inhabitants of these shadowy areas know well that state power can turn their territory into a 'black' space of control. Potentially capable to become a zone of transformation, the grey spaces are reflective of larger contradictions and conflicts in society (Yiftachel 2009).

Relatedly, AbdouMaliq Simone introduces the notions of the habitable and uninhabitable. In depicting how Kinshasa's green suburban boundaries exist, Simone highlights the *transitory* nature of the peripheral urban spaces (and of their nature): 'So the relationship between the habitable and uninhabitable oscillates, diverges and reconnects in ways that make the provision of 'new land, new opportunities, something that extends and builds upon the solidity of the existent city, but also seems to waste it at the same time.' There is always motion and commotion. And importantly, both the habitable and uninhabitable spaces are constituted by relations that test the integrity of territory and the knowability of the terrain of the city: 'The near universal perception in Kinshasa is that the city is increasingly moving elsewhere and, as a result, many inhabitants hurry to stake their claims at ever-shifting peripheries, which still seem to be in the middle of nowhere.' Simone continues:

> The rendering of the bush into concretized extensions of Kinshasa is of course in part driven by the "old standard" of escalating land values through speculation and the infusion of, in Kinshasa's case, external finance that jacks up property prices in older residential districts near the commercial core. (…)
> The extension of Kinshasa into its hinterlands prolongs a game that potentially runs out of space and time as the impacts of urbanization "talk back" through the shrinkage of virtuous terrain.[2]

From the green periphery of Kinshasa our eyes wander further east to the peri-urbanity of Gurgaon, the fast developing satellite of Delhi, India's capital (Gururani 2013). A recent article startles the reader with this opening sentence: 'The first thing you notice when you come to Gurgaon is the number of skyscrapers. The second is the pigs.... The villagers whose lands were bought and then developed, were squeezed into ghettoes near the high rises of Gurgaon. Their hogs and livestock wander freely through the maze of skyscrapers and private fences' (Doshi 2016). In contrast to the horizontal sprawl of Ontario and the creeping suburbanization of Kinshasa, Gurgaon's peri-urban landscape is in-your-face verticality that abuts starkly with the agricultural past and presence beyond. The grey zone and threshold in Gurgaon is characterized not just by contrasts between agricultural and urban societies, but also between rich and poor in dramatic fashion. Gururani reports that the suburbanism of Gurgaon is the product of a 'pastoral capitalism' where 'the discourse of green nature' has become a conspicuous referent of a 'good life' and where developers advertise their products with references to 'green pastoral settings' and 'lush green surroundings' (Gururani 2013).

From the beginning of human settlement, the relationships between town and countryside have been considered a central divide in our human experience. At the basis of much of the urban planning debate on urban boundaries historically, but especially right now, has been the need to use less land for more people that move to or move inside urban areas. We have learned to hold sprawl responsible for many of the urban problems that plague us today from global warming at the planetary scale to obesity at the scale of the body. We have also reached a professional consensus on curtailing sprawl with all manner of measures (Angel et al. 2010: 7). The belief in the importance of regulating urban form for the benefit of environmental sustainability and social justice has led to a strong consensus in the planning professions that dense is good and urban form can be a prime medium through which climate change can be affected if appropriate changes are made to the way we are building our cities. The normative consensus is built on some common sense convictions on the effects of density on energy, transportation, green house gas emissions and the like (Angel et al. 2010: 13). But researchers have cautioned against assuming that urban planning strategies for denser urban form might be universal in appeal and effect. One question that has been asked is: 'Could it be that urban containment and compact city strategies are now appropriate in some developed countries but inappropriate in many, if not most, developing countries?' (Angel et al. 2010: 7). The questions raised and partially answered, in the preceding sections on densities and boundaries point to a larger set of problematiques at the core of today's extended urbanization. They speak to nature in the age of the global suburb more generally. I will approach these problematiques in the final section of this chapter.

GLOBAL SUBURBANISM IN THE CAPITALOCENE

The suburbs (as a form, a certain idea and materiality of how cities grow) are identified as a major contributor to the Anthropocenic reality we find ourselves in at the beginning of the twenty-first century. Arboleda notes that '[t]he ecologies of the Anthropocene are, in sum, riddled with new ways of experiencing anxiety and distress and the planetary urban fetish becomes the more ruthless, the more remote the exploited and plundered ecosystems are from the point of consumption' (Arboleda 2015: 7). Yet, while there is no single cause or source for this fetishization, the production of urban space is a major process through which capital accumulation 'takes place' and through which it generates all manner of systemic inequalities as well as climate change-relevant perturbations. In urbanization, or in our context suburbanization, the destructive trajectory of the capitalocenic destruction finds an address and a culpable actor network of political economies that produce space and, at the same time, generalizes responsibility into an unrecognizable concoction of guilt that obfuscates the processes through which those political economies work. Mark Whitehead accordingly ponders 'the extent to which it may be helpful to consider our contemporary geological period not as the Anthropocene so much as the *Metropocene*: a period defined by the dynamics and demands of urbanization' (2014: 100; emphasis in the original). This argument receives nuance by differentiating between metropolitan and global scale impacts envisioned to be the consequence of urbanization. There is much space between those scales to differentiate along various class and other factors, various notions of impact and in fact, various impacted communities and places.

Speaking about the Anthropocene is, in the first instance, a look back at the impact of past practice. One question could be: how did we end up with a warming planet? Yet, T. J. Demos rejects 'the Anthropocene's terminological obfuscations and disavowals of culpability' (Demos 2015). This is less about general culpability and more about understanding specific causalities. This is not the species at fault but its involvement in processes of capital accumulation through particular spatial forms and ways of life. Through this, we are potentially leaving the 'black box' notions of 'industrialization, urbanization, population and so forth' behind and are also moving beyond a purely 'consequentialist' bias in viewing the period we live in (Moore 2014). So, we are moving the narrative from the consequences of Anthropocenic expansion to the modalities through which we built the world in which we live. If there are going to be ten billion humans on the earth, as one current projection holds, we know for sure that '[e]very last one of those ten billion human beings is going to need a place to live' (Greenfield 2016). Most of the future inhabitants of the earth's crust will be living in entirely new cities and many more of those who already live in cities will live in those new, mostly suburban environments, too.

What we call urbanization today is generally seen as having co-evolved with industrialization and the rates of urbanization rise sharply after the year 1800, which is also often seen as the starting date of what many consider the Anthropocene today (Crutzen 2005; Swyngedouw 2011; Zumelzu 2011). While Moore's definition of the capitalocene suggests that 'the origins of today's crisis lie in the epoch-making transformations of capital, power and nature that began in the 'long' sixteenth century' (Moore 2014: 1), industrial era urbanization is therefore largely coincident with what Moore may call the consequentialist definition of the Anthropocene (Barau and Ludin 2012). Yet it is exactly through the production of *urban* space typical of the industrial capitalist period that capital accumulates at the rates that have made it particularly aggressive in 'its extraordinary reshaping of global natures' (Moore 2014: 5). The 'great acceleration' of climate relevant changes from the Anthropocene occurs after the Second World War, a period which coincides with the Fordist-Keynesian regime of accumulation for which the 'suburban solution' represented a major regulatory mechanism for both economic crisis and social unrest (Steffen et al. 2007: 618; Walker 1977). The suburbs have carried the chief load in both conversations: they are considered failures in building sustainability and contributors to social inequality. Cities are seen as part of the solution and 'innovative response' (Hodson and Marvin 2010: 289). While suburbia and suburbanites are definitely considered to be among the chief 'causes', they rarely get included in the catalogue of the sites or origins for 'innovative responses' to the problem. They remain firmly on the debit side of the Anthropocenic equation.

Central to understanding extended urbanization is the question of land (Harris and Lehrer 2018) which also includes the topic of land grabs (Lazarus 2014). Sub/urbanization processes in some parts of the world have increased vulnerability to infectious disease outbreaks as has recently been the case in West Africa, where the peripheral extension of urban settlement has created inroads for the Ebola virus during the 2014 pandemic or in Brazil, where urbanization has been named as one possible cause of the rapid spread of the Zika virus in 2015 (Channel News Asia 2016). Another important global-scaled ecological consequence of suburbanization is the worldwide mining and use of sand for construction. If it is correct that much of the suburbanization of the mega-city peripheries in China, India, Turkey, the Middle East and elsewhere takes the form of high-rise construction in new towns at the urban edge, sand is a major ingredient. Concrete is largely made of sand. Dredging rivers and other bodies of water for the material is often done illegally and by so-called 'sand mafias'. It destroys local environments and livelihoods of established agricultural communities at a global scale (Beiser 2015).

In this context, there is consensus that cities are generally 'quilting points' (Swyngedouw 2011: 257) of the climate change conundrum. Yet, '[h]owever trendy "urban infill" and densification may be in our faculties of architecture, polities in the developed world are no longer willing to countenance the kind of conditions that inevitably result when hundreds of thousands of human bodies are packed into every square mile' (Greenfield 2016). Neither the dystopian refugee camps that can't be cities nor the utopian massive housing projects that grace the peripheries of emerging cities in Africa, Turkey, India and China that won't be 'urban' in the classical sense would be acceptable to the chorus of global northern density advocates. While urban form itself hardly carries with it specific ecological damages, the way it is delivered certainly does. Extending the speed and scale of Fordist-Keynesian or Soviet era massive suburbanization through innovative technologies such as the 3D printing of construction elements, 'China's world-historical rush of citymaking appears if anything to be accelerating' (Greenfield 2016). The Chinese government will hardly be able to stem the tide of fundamentally altering the capitalocenic impact of pouring concrete at such unprecedented pace into urban forms that are themselves hardly sustainable in any foreseeable future. If one adds the anticipated (and already experienced) rates of destruction at generational speed, the material ecologies and flow metabolisms involved are beyond anybody's imagination and far in extension of even the most brutal Corbusian excesses of the twentieth century.

SUB/URBAN POLITICAL ECOLOGY

Lawhon, Ernstson and Silver characterize the urbanization of nature as 'the social, cultural and political relations through which material and biophysical entities become transformed in the making of often unequal cities' (2014: 500). This includes stretching the notion of extended urbanization to the mines and fields in all corners of the earth from where concentrated forms of urbanization (i. e. cities and suburbs) are being provisioned (Arboleda 2015: 7). The suburbs as a place of alienation and separation from nature, therefore, are to be understood themselves as a place of a larger chain of activities that constitute the extended urban political ecology. But, as I argue here, the suburbs can only be subsumed partially under an analysis that foregrounds a generalized urban that ostensibly rests on particular centralities as its point of origin. As Merrifield (2013: 15) has so convincingly argued: 'It's a new reality, the result of a push-pull effect, a vicious dialectic of dispossession, sucking people into the city while spitting others out of the gentrifying center, forcing poor old-timers and vulnerable newcomers to embrace each other out on the periphery, out on assorted zones of social marginalization, out on the global *banlieue* – left to fend for themselves on *world market street*'.

The suburban is understood here in two ways. At one level, suburbanization is a process of 'massive' city building and re-building at the metropolitan edge including processes of post-suburbanization (Guney, Keil and Ucoglu forthcoming). At a second level, we are dealing with new everyday suburbanisms, new suburban ways of life that are central to understanding the link of planetary sub/urbanization and the Anthropocene. It is in the everyday behaviour of the suburbanites that the 'suburban problem' finds its most common 'situated' expression (Lawhon, Ernstson and Silver 2014: 506). For the people of Huasco, a Chilean mining town from which far-flung parts of the world are being provisioned with the material commodities of growth, everydayness is fraught with death and disease (Arboleda 2015: 12). While the everydayness of the suburban is a place where metabolic rifts become visible in the environment of the 'world-ecological uncanny', there is much variety in the field of extended urbanization. The suburban silicon valleys and alleys in the Global North are, in fact, connected intimately to 'the socio-ecologically dystopian geographies of Mumbai's or Dhaka's informal suburban wastelands' (Swyngedouw and Kaika 2014: 463). Suburbs both, but of a different kind. This is part of what Roberto Monte-Mor has called 'the combined process of metropolization and extended urbanization' (2014a: 112). The suburbs, major part of the operational landscapes of the extended urban society, are not inhabited by immobilized suburbanites whose only political reflex is to defend their privileges. The question of extended urbanization is hence both spatial and political (Monte-Mor 2014b: 265). It underlines the domain of suburban political ecologies.

CONCLUSION

Looking at the suburban political ecologies in this more complex way allows a view beyond Lefebvre's habitat towards the suburbs as contradictory terrain with contested densities and boundaries. They are also a site and terrain of political ecological action. An important aspect of this problematique is the debate on density and infrastructures. We can posit that between the exostructures of the global urban political ecologies that sustain both Huasco and the suburbs of Toronto (for example) and the people as infrastructures (Tonkiss, 2013; Simone 2004), there is real life, politics and potential for change. The suburban is both provisioned by extended urbanizations (such as the kind Arboleda describes for Huasco) and it is extended urbanization. This conceptual unity is lost in common depictions of the suburban as a net winner (and hence culprit in) Anthropocenic excess. Its boundaries are always regional and transnational. The suburban is not a peripheral or marginal place (or topic). Large majorities of urban society now claim the suburban as home, workplace or landscape of *jouissance*. The relevance of these landscapes for today's urban political ecologies need to be one starting point for the performativity of urban and social change (perhaps radical incrementalism of sorts, see Lawhon et al. 2014: 510). Arboleda is correct in drawing our attention to the 'operational landscapes' of Huasco which he calls 'a metabolic vehicle of planetary urbanization' (2015). That also includes Fort McMurray, 'the suburb at the end of the highway', in the Alberta oil patch (Major 2013). Yet, just as mining and logging sites are part of the invisible landscape of planetary urbanization, the suburbs with their dumps and warehouses, their land uses that have been expelled from the inner city, are, too. Everyone sees the airport terminals but nobody sees the aerotropolis. Those aspects of the suburban remain part of the 'urban uncanny' (Kaika 2015). A political ecology of the Anthropocene might have to be reimagined in those global suburbs that now house the majority of us and from where the capitalocene derives its metabolic dynamism.

Anting New Town, China

9 The Political Suburb

When Donald Trump was elected President of the United States on 8 November 2016 many reasons were given for this surprising result. Among the constants in the analysis was the realization that while he did not win a single big city, the suburban and rural vote carried him to victory. Florida can be seen as an example. In that state whose delegates to the Electoral College put Trump in a solid lead on election night the diverse cities supported Hillary Clinton while the 'suburban white vote' bolstered Trump's win (Smith and Bedi 2016; see also Myerson 2017). In fact, whether it is California's 'anomic suburbs' of the 1960s that were the breeding ground of the John Birch Society, Proposition 13 and Ronald Reagan or the petrochemical suburbs of Louisiana that are the home of Tea Party type anti-governmentalism, the white urban periphery has long and rightfully been suspect to political commentators (Hochschild 2016a, b). Resorting to this standard analysis of a conservative white political community on the periphery may be surprising in light of the changes and upheavals I have discussed in the preceding pages but the residue of the suburbs as a white space in this manner is certainly a reality of note. Much more, though, in this discourse, the suburb has become a 'state-of-mind', a generic political mindset, a form of existence that can exist anywhere, not just in the actual geographical periphery. An early report on election night marked this space in the following words: 'Staten Island is the city's version of white suburbia, a microcosm of the half of America that cast its ballot for Donald Trump. Its politics are quiet and suburban; its bigotry is quiet and suburban, too. You might not have realized how sympathetic its residents were to cops until the investigation of Eric Garner's death-by-chokehold a few years ago, or that they believe that 'the Mexicans get drunk and start trouble' until a local paper went right out and asked' (Osberg 2016). Somewhere between the declining American industrial middle and the 'rugged entitlement' of the new American periphery (Heiman 2015), we find the panoply of new neoliberal subjectivities that are easily labelled as 'suburban' (Peck 2015a; Nijman and Clery 2015). The suburban as a real and imagined space of politics and governance has thus come to take up a central role in the narrative of the crisis of the current political world, not just in the United States but similarly, if also always differently, in the UK in a time of Brexit, France facing a surging Front National and Brazil in a post-Workers Party political universe.

The suburbs have long been seen as a problematic space in this fashion. As early as 1994, Mike Davis wrote about Los Angeles politics: 'The failure of candidates to address, or even grasp, the acuity of the suburban malaise explains...much of the populist rage that currently threatens the two-party status quo. America seems to be unraveling in its traditional moral center: suburbia' (Davis 1994). The suburban as a trope for general (political) malaise is just one side of the political imagery recently associated with the suburban. Revisiting the Los Angeles sub- and exurbs almost 20 years after his early 1990s observations, Davis (2011) subsequently discovers, in the Imperial Valley ravaged by unemployment and poverty, a rebellious form of organizing not unlike the Occupy protests that had sprung up across cities around the world. The alleged centrality of protest – reconfirmed by the protests associated with mass eruptions on central squares in cities since the era of the Arab Spring and the struggles against neoliberal hegemony in Southern Europe (Sharp and Panetta 2016; Dikeç and Swyngedouw 2017; Kaika and Karaliotas 2016) – somewhat obscured the view of a politics beyond the centre. The central location of Tahrir Square, Syntagma Square and Gezi Park became synonymous with the centrality of politics, even though in Cairo or Sao Paulo, for example, the city centre was just used as a temporary stage for a politics that emerged from the periphery. Not all of the peripheral urbanization is actually in the geographical margins of the urban world but such developments clearly change the politics of the city overall (Caldeira 2016).

We must now realize that our urban political world has grown beyond the centres. When Lefebvre coined the idea of the 'right to the city' as a reaction to the exclusion of the Nanterre students from the promise of Paris, he did not endorse the fetishization of the 'decision-making centres' (2003: 119). Rather, he cast a light on the subordination of the peripheries where the '99 per cent' of *his* era of late Fordism worked, studied and mostly lived. There has been a Haussmannian history of evicting the working class from the centre that structured the political arena in that city and even in the country of France. What's more, the socio-spatial dialectics that has characterized politics in Paris is now a ubiquitous presence in the gentrified, corporatized capitals of capital from London to New York, Istanbul and Toronto where poverty and diversity are evicted from the core. Perhaps there is a strong lesson here: we might have to increasingly disregard the symbolic but ultimately vacuous power centres of global capital and concentrate on the politics of everyday life where it actually goes on. This might just be in the suburbs and inbetween spaces of our metropolitan areas. From there, we may hear the battle cry: 'occupy the strip malls' instead of squatting the traditional 'decision-making centres'.

SUBURBAN GOVERNANCE

Historically, suburbanization has been regulated through three modalities: state policy, capitalist accumulation and authoritarian forms of privatism (Ekers, Hamel, Keil 2012). The tension among these modalities is characteristic of the way in which we imagine the political metropolis today as a new neoliberal era of city-making has once again re-calibrated the role of planning and markets and their relationships to each other (Boudreau 2016; Fontenot 2015; Kohn 2016; Magnusson 2011). The ongoing and one might argue accelerating proliferation of urban society necessitates thinking anew about urban politics and governance. This claim is not entirely new and it doesn't stand alone. A rich comparative literature on urban governance, regional governance, new urban politics, etc. exists (for a partial review see Keil et al. 2016). The mainstream urban literature in the United States and Canada has long claimed that metropolitan areas are getting shortchanged in the federal political system both constitutionally and in everyday political processes despite their overwhelming political economic presence (Katz and Bradley 2013). European scholars for their part have long talked about 'collective agency' for cities and regions (Le Galès 2002). The idea of collective agency also rests on the tapestry of state and non-state actors from which urban and regional decision-making draws its strength. It is commonly assumed that in North America and the UK government and business coalitions form 'regimes' which also include, at times, strong contingents of civic organizations (Stone and Sanders 1987). Yet, in contrast to European urban, metropolitan and regional development, American 'growth machines' are tightly controlled by interests that tie capital accumulation to land development (Logan and Molotch 1987). In the past, development politics was thought to have taken place in mostly two locations: (1) the redevelopment of the inner city – previously through large-scale government programmes such as urban renewal and more recently through gentrification, also often state-led and (2) suburban development where mostly private actors build cookie-cutter subdivisions on green fields. This has traditionally included residential tract housing as well as shopping centres, infrastructure and industrial land uses. Regime theory has done a marvellous service to the understanding of urban agency as it rejected the facile assumptions of pluralist power sharing and elitist conspiracies.

The opening up of our understanding of urban and regional government to more than elite actors and shifting interest groups but to see a systemic political economy at the core of urban and regional decision-making meant the acceptance of new political spaces between, as Hartmut Häußermann called it, the 'big city of coldness' and the 'big city of warm nests' (cited in Farias and Stemmler 2012: 62). We need to ask: what kind of politics springs from this difference? For our purposes here, this implies the question of how exactly the three modalities of suburban governance – the state, capital accumulation and private authoritarianism – are intertwined. The idea that we need metropolitan governance is not new. In fact, it really is born together with the suburbanization process after the Second World War. The metropolis as an outcome of Fordist-style industrial capitalism-cum-automobility needed territorial regulation. Eventually, the real development of centre-to-periphery expansion was abandoned and replaced by the view put forward most forcefully by the Los Angeles School, that the governance of regions had to be rethought more fundamentally as centres (from where governance used to be imagined conventionally) started to lose their gravitational force in metropolitan regions like Southern California (Dear and Dahmann 2011; Mayer 2012; 2017). The regional city is not as bounded as classical, territorially oriented political science has it, but also not as boundless as poststructural researchers tell us it is. The region is constituted through a set of discourses, territorial boundary-setting exercises and technologies (Addie and Keil 2015).

FRAMING THE MECHANISMS OF SUBURBAN GOVERNANCE

Some suburbs are built overnight, some are the product of long-term processes. None of them happen naturally. They all follow some modalities of governance. As one recent observer of a century of American suburbanization observed: 'Many people – particularly planners – look at the American landscape and see disorganization.... As an engineer, I don't see disorganization. In fact, I see one of the most highly organized mass endeavors ever undertaken by humanity. We have transformed an entire continent around a new theory of development. This required incredible levels of centralized coordination on policy, finance and regulation. American development is exquisitely organized' (Marohn 2016).

Marohn, who is not shy about his own agenda of returning city building to the principles of a pre-automobile age, refers to what he calls 'America's suburban experiment' (Marohn 2015), by which he means building a landscape where 'new growth happens on a large scale and... construction is done to a finished state; there is no further growth anticipated after the initial construction' (Marohn 2016). We may debate Marohn on how 'large scale' and how 'finished' those landscapes may have ended up being but it is clear that they were the product of a sophisticated set of rules and that they were delivered through joint efforts, in most cases, of individuals and communities, states and markets at various scales.

Marohn notes the complexity of the 'suburban experiment' where nothing was left to chance (Marohn 2015). In all that, infrastructure was often 'the message' or a *medium* through which location, form, morphology, density and other factors were determined (Filion 2013b). To Keller Easterling (2014: 73), as we learned in Chapter 7 above, those suburban infrastructures themselves are more than a medium in the end, they are a *disposition*, they have agency, 'extrastatecraft', a form of governance with 'political chemistries and temperaments of aggression, submission, or violence – hiding in the folds of infrastructure space'. The suburbs play a particular role in the governance of this space. The urban periphery defined as 'a field of mass-produced suburban houses is a common phenomenon in infrastructure space and it is an organization with clear markers of disposition' (Ibid.).

What Marohn calls 'experimental' is dubbed outright 'parasitic' by Robert Beauregard (2006) in his narrative on 'when America became suburban'. By parasitic, he refers to the fragmented and non-redistributive nature of suburbanization. But Beauregard also points to the fact that with the peripheral growth 'urbanization had jumped to the metropolitan scale' and hence demanded a new mode of governance that was different from that by which centralized, redistributive cities were governed in the past (Beauregard 2006: 4). The parasitic suburbs were not just part of a failing system of metropolitan governance in which central cities were left to die in the midst of prosperity, their dispersed landscape was also a passive element in the Cold War defence against nuclear attack and in this sense their governance was guided by more-than-regional concerns (Beauregard 2006: 147–55). More explicitly, as Andrew Friedman has shown, suburbia played an active role in the landscape of American imperialism as exemplified by the 'covert capital' spreading through the Dulles Corridor in suburban Fairfax County west of Washington D. C (Friedman 2013).

But the United States and its Anglo-Saxon cousins (Australia, Canada, the UK) were not alone in building large-scale and readymade, automobile suburban subdivisions. Europe, most notably, had its own version of the 'suburban experiment' although, admittedly, planners and residents there saw themselves often as builders of city extensions, rather than refuges from the city. In Europe, there has been a range of suburban development in post-socialist states and even within Western Europe and North America, where we see denser or non-conforming forms of suburbanization (Charmes and Keil 2015). In Western Europe, for example, suburban settlements, often built as so-called 'satellite cities', became the very expression of modernity and centralized governance. The large-scale housing estate on the periphery enabled 'a consolidated social welfare state to demonstrate what it was able to do' (Harlander 2011: 18; Phelps and Vento 2015). In turn, as Hirt and Kovachev (2015) have observed, 'large cities across Eastern Europe were, in the span of about two decades, surrounded by massive pre-made-panel housing blocks with a grandiose but Spartan flavour' (Hirt and Kovachev 2015:

181). These housing estates were the product of long-term planning linked to industrialization and organized by the central state (Logan forthcoming).

In all cases, governance of suburbanization as a process of building and re-building urban peripheries as well as governance of suburbanism as the regulation of everyday life was central. Those European cases of modernizing explosion, for example, made Henri Lefebvre rethink the urban question more generally (Lefebvre and Ross 2015).

Further afield from – but in no way secondary to – the Anglo-Saxon or European experiences, we note the importance of governance in suburban expansion in the Global South. From when the first squatters of Morro da Favela above Rio de Janeiro or the first Anatolian migrants in the *gecekondu* of Istanbul pitched their shacks overnight, even the most tentative or fleeting of informal settlements on the periphery followed some guideline of development that paid tribute to the local circumstances. The claiming of space on the periphery of the city has always been an intricate dance whose steps are part condition of land tenure and legal rules created by the authorities and part free improvisation, sly moves around the status quo.

One of the key points of agreement in the blossoming literature on suburbs is that diversity is the norm rather than the exception, which means that Levittown is but one form of suburbanization and it forces us to approach them within urban theories that reach 'beyond the west' (Edensor and Jayne 2012). In terms of the governance of suburbanization, Ananya Roy (2009) pushes us to consider multiple worldly forms of governance, not as derivative of the US experience, but rather as central to the increasing suburbanization of urban-regions in all spaces including the US.

The question of formality and informality in the governance of suburbia is ever present. Even the least fixed of peripheral settlements in the United States, the nation's mobile home parks where twenty million people live in 8.6 million units, are strictly regulated by codes of construction and conduct (Bréville 2016). This is also true for the persistent mega-developments of post-suburban landscapes that are indicative of a generalized risk society: a planned grid of interconnectivities of hard infrastructures and soft everyday living in which governance emerges as a constant in/formal *bricolage* of sorts (Phelps 2015). Layered forms of informality, squatting, entrepreneurialism and state action characterize many sub/urban expansions such as those that Coker (2016) and Ortega (2016) observe in Metro Manila.

MANIFESTO OF THE SUBURBAN REVOLUTION

Urban society is not a virtual object anymore as it has materialized many ways that were still emergent when Lefebvre formulated his original theory. Yet it has also obliterated our conceptual categories, often beyond recognition. There are no essential differences anymore between centres and suburbs. The politics of everyday life becomes central to the new materiality of the urban periphery. We can therefore also speak about the right to the suburbs (Carpio et al. 2011). There is no more catching up to do. There is no deficit that has to be overcome: 'Within the postmetropolis, a reconfiguration of the meaning of urbanism is taking place. Centrality is increasingly reserved for immaterial networks of power and the physical assets that support them, while bodily existence within the postmetropolis is increasingly moved to the periphery' (Quinby 2011: 75).

We end up with a web (or we could say assemblages) of post-suburbanization processes. In their totality, they add up to a suburban involution. We witness not a complete dissolution but a reconsolidation of the urban fabric, even a balancing and in any case a rejection of classical functional or conceptual dichotomies such as live-work. The process of post-suburbanization entails a profound re-scaling of the relationalities and modes of governance that have traditionally regulated the relationships between centre and periphery in the suburban model

(Phelps and Wood 2011; Phelps and Wu 2011; Hamel and Keil 2015). The '(post)-suburban involution' has its own 'functional, spatial and rhythmical diversification of post-suburbanisms' (Charmes and Keil 2015); we are talking about 'a complexification and folding in of suburbanizing cultures and rationalities, as opposed to a linear process of centrifugal succession' (Peck, Siemiatycki and Wyly 2014: 389).

The culture of politics in post-suburbia is now very self-conscious of the notions that come with density in the metropolitan arena. The periphery begins to reveal a different set of political options than the one to which it has historically been tied. In her work on the governance of the postcolonial suburb, Roy (2015: 344) has pointed out that 'the politics of the suburban periphery is the politics of political society' through which emerging contradictions of centrality/peripherality, formality/informality, state/market are renegotiated. In this sense, this thinking contributes to a general 'critique of stable categories of space, society and state' (Roy 2015: 345).

Suburban politics are strongly influenced by the density debate (see the previous chapter). In the past these politics have been most directly associated with the privatism of suburban households and communities. Perhaps the clearest expression of this relationship has been the liberal or neoliberal tradition of public choice as the regulating principle of politics in the fragmented peripheries of America's metropolitan peripheries. What was sold as a politics of individual rights and opportunity meant, in reality, 'privatizing with class' (Hoch 1985), enclaving at best and segregation at worst. As we re-evaluate the role of the suburban in these current high times of neoliberalism, we can follow Jamie Peck's assessment that, indeed, 'American suburbia has been relationally defined, in ideological terms, as the dispersed other of metropolitan Keynesianism and in social terms as a haven from both big cities and "big-city problems"' (2015: 130). In the US context, but also elsewhere, this 'dispersed' form has also often been associated with low density morphologies as landscapes of freedom's desire, whereas 'towers in the park' or related mixed forms of suburban densities were either lumped in with the 'Keynesian' metropolitan project (as it happened often in the writing of the histories of Toronto's 'city that works' of double-tier metropolitan governance, part of which included the peripheralization of higher densities out of the Victorian low density core). Those islands of high density in the periphery have forever been written into a narrative of modernist and planning failure. The dense/city versus dispersed/suburb trope has been an ideal carrier of the urbanist differentiation of form and function: it became, to some degree, the ideal battle ground for ideas of human life that are entirely unrelated to how we are housed, sheltered and moved around.

The question of density in politics often sidesteps the real issues metropolitan regions nowadays must confront: how to create sustainable environments that are just in terms of social distribution and access to resources such as mobility, education, health care and jobs. It doesn't really matter, in this context, whether the ultra-rich, for example, live in penthouses around Central Park in New York, along the Thames in the core of London or in a 'giga-mansion' on shaved hilltops in Los Angeles. The density of their exclusivity and the location of their residences are utterly unimportant compared to the political power their inhabitants and their class exert over the entire region. At the scale of the metropolitan region, then, politics is beginning to give rise to new discourses that bypass the dead-end road of density dichotomism (Haldeman 2015; Story and Saul 2015). The reality of hyper density that changes the reality of real estate markets and politics in large internationalizing cities stands in opposition to the powerlessness and poverty in dense tower neighbourhoods at urban peripheries from Glasgow to Paris and Leipzig to Toronto. The giga-suburbanization and vulgar, gated neoliberal mansions (Knox 2008) that sprawl into the outskirts of many urban regions around the world, from Los Angeles to Shanghai, for their part, stand in opposition to the equally sprawling bungalow suburbs that get built in their shadow as well. Density per se does not provide a sharp lens to predict or understand the type of urbanism that takes shape and the politics that naturalize its polarities into winning and losing modes of urban life.

Much of the politics of the new periphery is shaped by path-dependent institutional arrangements and existing territorial governance processes. In Eastern Europe, for example, Stanilov (2007: 187) has observed that '[m]any of the new suburban communities, popping up beyond the concrete walls of communist housing estates, still lie within the administrative boundaries of the metropolitan jurisdiction'. Meanwhile, in France, Touati-Morel (2015) has shown 'hard' and 'soft' densification policies in the Paris city region that reflect the different socio-spatialities and variable densities of municipalities there. For their part, Poppe and Young (2015) explain how the jurisdictional proximity of inner-suburban low and high-rise neighbourhoods lead to a particularly complex set of political imbrications in Toronto's efforts to accomplish 'renewal' of that city's tower neighbourhoods.

THE URGE TO RETROFIT

Much of the politics and governance of the suburban today relates to the idea of 'fixing the suburbs'. The literature on suburban 'fixing' is, however, quite distinct from the extensively growing overall work on urban retrofitting where suburbs are often just an afterthought (Bouzarovski 2016; Hodson and Marvin 2016). Everyone seems to agree that the suburbs – too often lacking in physical and social diversity, people-oriented places and travel alternatives to the car – need some work. Given that 'suburban constellations' can now be found in the booming peripheries of cities from Cape Town to Shanghai, from Toronto to Montpellier and from Santiago to Helsinki, the question can reasonably be asked whether the retrofitting of suburbs is now 'an American or a global trend' (Dunham-Jones 2015)?

One thing the literature (and practice) of suburban retrofitting is certainly accomplishing is that it carefully and persistently contributes to de-mystifying and de-stereotyping suburban life and form. In fact, there is now a pervasive realization, to borrow a phrase from Fran Tonkiss, that any urban *form* just has a 'temporal life' (2013: 56). And moreover, the places where one finds urban *life* are often unpredictable and mostly unconnected to urban form. Influenced by Saskia Sassen's idea of 'cityness' as a space of encounters, Tonkiss notes that those encounters may happen in places that have not necessarily been originally designed to host them. The suburban strip-mall may be the best example for this. Designed for easy and largely anonymous access by automobile, for a decluttered, individualized use of space, they have now become the epitome of immigrant, ethnic interactive economies in ethnoburbs and beyond, the economic wellsprings of arrival city economies (Saunders 2011).

In a world, though, where 'urbanity as a form of life has emancipated itself from the cities and has long nested in the urban hinterland, even in so-called rural areas' (Bormann et al. 2015), we might speak of a phenomenon of sub/urbanity instead. The process of generalization of sub/urbanity as a form of life works both ways: just as much as we now find stylized plazas and 'green rooms' in the lifestyle malls of exurban communities, the proximity to 'landscape', a traditionally non-urban trope, is now part of the definition of cityness itself (Bormann et al. 2015: 114–15; translation by RK). Retrofitting suburbia then operates on the assumption that bringing urban morphological diversity and urbanity into the suburbs will somehow bring improved everyday functionality, visual pleasure and vibrancy. But for retrofitting thus defined to truly be transformative and effective, it must first accept that suburbia itself is not monotonous but consists of various partial spaces that are ripe for change to reflect the demands of climate change adaptation, economic crisis management and societal shifts in demographic and family structure. Lastly, when we talk about a politics of retrofitting suburbia, it would need to include all of its constitutive landscapes of housing, commercial and office or manufacturing spaces (Jessen and Roost 2015b). Yet, in reality, suburban landscapes, although they were the product of an unprecedented subsidy of general taxpayers to homeowners through housing loans and infrastructure construction, were ostensibly domains of privacy. This has had well-documented consequences for their sociality (or perhaps lack of it) and a certain leaning towards regional isolationism and a distinct drawbridge mentality. This point needs no further exposition here but in our context it has consequences. The suburban single-family home and its privatist governmentality stands in tense contradiction to the modalities of government intervention and accumulation strategies. Together these modalities have produced a landscape without apparent guidance in planning terms and without authority to do something about its obvious shortcomings (Bormann et al. 2015). At the same time, though, public domains in suburbia have become targets and arenas of a politics – such as transportation, housing and open space – of retrofitting that have become central to the urban region overall.

THE CHANGING SCALE OF SUBURBAN PLANNING AND POLITICS

For the emerging politics of the post-suburban landscape, it is important to note that once the Tieboutian scale has been jumped, all manner of spatial reorientations in the region and beyond are possible. De Meyer et al. (1999: 31) have noted '[t]he morphology of post-suburbia, which should be viewed at a wider regional scale, evinces more similarities with the concept of "megalopolis" as introduced by the French geographer Jean Gottmann'. In this changing geography, scale is an important marker of both state order and critical possibility for struggles around 'the right to the suburb' in inner (Carpio et al. 2011) or outer (Schafran et al. 2013) peripheries.

The political landscape we now encounter in post-suburbia is linked to morphological change which, in turn has its roots in larger scale changes to socio-economic and demographic restructuring in metro regions overall. Nico Larco (2010: 70) has found that '[t]he stereotypical suburban image of the single-family home and the nuclear family is no longer the exclusive reality of suburbia. Maintaining this image impedes our ability to create policies that engage the actual composition of suburbs.' The political consequences of this disjointed perception are now beginning to be felt in both apparent mismatches of policy to the actual terrain of post-suburbia and in the emerging political reactions as a new set of demands express the growing awareness of social difference and morphological variation in the metropolitan periphery. Multi-family housing, for example, appears to become more widespread in the periphery more generally. There are almost ten million such units in American suburbs alone. They come in three types: condominiums, elderly housing and 'mixed-use lifestyle centres' mostly in the elite income bracket. Suburban multi-unit housing is also characterized by a significantly different demographic/family structure as it is home to more singles than the classical subdivision (Larco 2010: 70–5). In areas where smart growth rules are in place as is the case in the province of Ontario, there seems to be reason to believe that despite only modest progress towards higher density development region-wide, the discourse and the technologies of regionalism have shifted towards a redefinition of the territory (Addie and Keil 2015). New urbanist developments may also contribute to more variety in the suburbs although critics warn that 'the development [of new urbanist communities] may also increase sprawl by inducing households to leave their high-rise developments' (Skaburskis 2006, quoted in Fiedler and Addie 2008: 15).

The discursive shift due to changing socio-spatialities, densities and morphologies can be observed in a variety of settings. Fleischer (2010), for example, has noted that ideological amalgams of modernity, environmentalism, middle classism and consumerism have now become effective in suburban conversations in Beijing and elsewhere in China. The traditional (Maoist) productivist orientation in ordering space was given up in favour of a consumerist orientation and a dominance of the middle class in sub/urban political and social spheres. Everyday life plays a huge role as a conduit and terrain of social differentiation. At the same time, '[m]iddle-class homeowners are increasingly aware of and willing to claim their growing power and agency in shaping the suburban space' (Fleischer 2010: 149). Drummond and Labbé (2013) have made us aware of very similar tendencies in the Global South where the suburban experience is increasingly characterized by multiple simultaneous transformations that will require a close look at actual social interactions in these changing morphological and social constellations. While the circumstances of the production (and consumption) of space in postuburban China and the Global South are different from the classical case of US suburbanization, the changing morphologies are related, as they are in the West, to a more general shift in how the postmetropolis takes shape. This development echoes and simultaneously qualifies, the prescient observations by Henri Lefebvre on the transformations that characterized 'extended urbanization' more generally. Critiquing the ideal of 'Unitary Urbanism' that was identified in the 1960s with Situationist ideas on urban change, he dated the end of city-centred urbanism with 'the moment that the historic city exploded into peripheries, suburbs – like what happened in Paris and in all sorts of places, Los Angeles, San Francisco, wild extensions of the city' (Lefebvre and Ross 2015: 49).

CHANGING POLITICAL ECOLOGIES OF SUBURBANIZATION

The extended (suburbanized) region is the basis for capitalist production, a scale of cognition and a terrain for action (Keil et al. 2017). In the context of this book, this has two immediate consequences. First, suburbanization was presented as one of the defining processes of our times. The suburbs entail capital switching at the cost of both human life and ecological devastation (climate change, etc.). Capital switching involves the redirection of capital accumulation from the production process into circulation and related spheres of consumption such as the production of space, the production of nature, education and so forth (Harvey 1982). The virtuous cycles of investment embedded in suburbanization seemed self-propelling for about half a century in the Fordist West, especially in the USA. They are now repackaged as recipes for entire national economies in countries such as Turkey, India or China (Ucoglu 2016); and they come, at first glance, with a package of depoliticization which is exemplary as it connects key features of indebtedness, dependence on the horizontality of organized consumption, extended metropolitan connectivities (and disconnectivities) with isolation, control and disjunction from the political in the conventional sense. Secondly, I am interested in the political ecologies of particular boundary settings that come with suburbanization. Although urbanism now tends to emphasize the positive environmental effects of creative economies in central (globalized, gentrified, normalized) cores, some of the most dynamic changes in today's urban world occur at the cities' peripheries (Ranganathan 2014; Ranganathan and Balacz 2015).

This leads to two related thoughts: First, those peripheries and their boundaries deserve more study. We know little about how the new sub/urbanites engage the metabolisms of their everyday lives beyond caricatures of suburban lifestyles that are bandied about uncritically and often in blatantly reactionary ways by urban-centred political ecologists. We need better theoretical and empirical work to help us understand how everyday suburbanisms function and contribute to the urban world in which we live. Secondly and relatedly, this leads to positing that the urban political ecologies of suburbanization in a global urban world are not static, conservative and ahistorical but that they are, in many ways, the hotbeds of new political ruptures and eruptions themselves.

Common assumptions in critical urban studies foreground the central place as the hub of metabolisms and the location of relevant politics. The last mile, premium network spaces, financial institutions bankrolling the water system, knowledge production, etc. all represent urban politics as spatially central. In terms of performing the political, central places, urban centres are privileged as locations for meaningful disruptions. Instead, perhaps, it may be useful to expand such political geographies to 'the street', the mall, the suburban neighbourhood as places aspiring to centrality. This would call for taking into account particular performative politics by often overlooked polities/constituencies that are involved in rapidly emerging metabolisms at the cities' (and often at the world's) edge. The mining town of Huasco, like Ferguson may, then, be everywhere, not just sequestered in a place of eternal marginality (where, in the case of Ferguson and similar suburbs experiencing impoverishment, the reflex is to use the same marginalizing language that has, in the past, been used to speak about 'the ghetto' and hence to relegate the political in these spaces to the margins, too).

The suburban makes us ask questions about the region, more than the neighbourhood, as foundation for collective life. Suburbia is a conflicted landscape of enjoyment (Lefebvre 2014) and it is part of Lefebvre's dreaded 'habitat', an alienated product of complex modalities of governance between the state, capital accumulation and private authoritarianism. Gated communities and *banlieues* are two sides of the same twisted coin. They add up to the 'negative utopia' of the 'bureaucratic society of controlled consumption' (Lefebvre 2003) and the aspirational 'arrival city' of the sub/urban century (Saunders 2011). The range of everyday political ecologies is defined broadly through the suburban 'explosion' between aspirations of middle-class life and a generalized 'architecture of enjoyment' and the camps and squatter settlements at the outskirts of many cities around the globe today. The suburban everywhere creates its own boundaries as it expands. At the same time, the suburban has consistently been a wellspring of activism and resistance (Mayer 2017; Schafran et al. 2013).

Planetary suburbanization creates a threat to humanity (Lefebvre 2014: 569). This poses new political priorities: suburbanization, not urbanization now leads to new imperatives. Still, perhaps the spatial imaginary at the basis of our political imagination needs to be shaken up more profoundly. The planetary urbanization patterns of today may not be the only ones we can imagine in the future. We may have to think more of fields than of towers (Ajl 2014: 541 and 546). Rather than hitching our political imaginaries to the extremes of hyperdensity and supersprawl, it is possible and much more plausible to imagine an entirely urbanized world society in a variety of morphologies and densities from which spring a plurality of tantalizing political possibilities in and beyond the Capitalocene. This may also give us a glimpse of the performative politics of the (de-centralized) everyday and of the (central) revolt at the edge of the suburbanization we witness today.

Such a horizontalized politics also needs some rethinking of the dialectics of spectacular (or operatic) and everyday politics. Lingering processes of 'roll-with-it'-neoliberalization are here to stay as that particular social formation fights bloody and prolonged rear-guard battles in the suburbanizing societies in which we live. Importantly, this re-situates neoliberalization from a process of realizing a Western master-discourse to a set of co-produced practices, inter-references and de-centralized engagements with fragments of travelling theory (Robinson and Parnell 2011; Roy 2011). We are now experiencing political challenges at the sub/urban everyday that are at once local, regional and global (Monte-Mor 2014a). The geography of political performativity has moved to the periphery where urban growth (and sometimes shrinkage) dynamically changes the terrain of the political demographically, economically, socially, culturally and environmentally. This is true for the traditional middle-class suburb as much as for the rapidly expanding poor peripheries of expanding Southern metropolises, the self-built, 'informal' urban peripheries and squatter settlements, as well as the in-between post-suburban landscapes that have become the majority habitat in many (eastern and western) European and American cities, or the new towns in East Asia that mirror and perfect the modernist suburban tower neighbourhoods of the second half of the twentieth century, themselves anticipating their post-suburban moment in a generation's time. The shift in focus inherent in this third dimension entails also a change in focusing on coming to terms with the specific challenges posed by the periphery to the climate change problematic. Put differently, this means that while the focus is now on the periphery as a problem to be solved by ideas and actions from the *centre*, the periphery takes its fate into its own hands by accepting, realistically, that if we can't address the problems of the most unsustainable places, the sprawling expanses of our suburbanizing world (Ross 2011), we have no future at all.

Waterloo, Ontario, Canada

Notes

1 Introduction
1 (http://www.theguardian.com/cities/2016/jan/28/where-world-newest-cities-look-same)

2 Suburbanization Explained
1 The original French formulation is slightly, but perhaps significantly different, as it does not use the verb 'belong' and speaks of 'populaire' instead of workers (as is common in French): 'Il y eut une époque ou le centre des villes était actif et productif, donc populaire' (Lefebvre 1989).
2 In addition, as Gibson et al. (2012) have pointed out, creativity needs to be seen less as a uniformly placed quality and more as 'relationally situated and linked across all parts of the city' (p. 287). Part of this relational landscape can be 'beachside suburbs with unique cultural histories' (p. 287) and places with 'de-centralized' small-scale cultural infrastructure…are highly valued and perform an important function in encouraging vitality and creativity' (p. 299). See also Flew et al. 2012.

5 From Lakewood to Ferguson
1 The 'homogeneity' angle is obviously overplayed here for effect. Herbert Gans (1967) among others already noted the diversity of suburbia in his classic study of Levittown, a suburban settlement quite like Lakewood.
2 Anatomy of Los Angeles, 1969 https://www.youtube.com/watch?v=7-R1b2Tz9fY ('Filmed as part of the French television programme Point contrepoint this critical and poetic portrait of Los Angeles, a sprawling city where the automobile is king and millions of individuals live together but never meet. During the documentary students, a French chef, authors Henry Miller, Norman Mailer recall the positive and negative aspects of the city.')

3 Importantly, though, some have pointed out that poverty in inner cities is still more dire than in even the most marginalized suburban communities of colour (Lewyn 2016).
4 These were both concepts deployed with much public fanfare by the United Way of Toronto to point towards the rising social needs of the inner suburbs of that City (United Way 2011).
5 Admittedly, though, even in those core Fordist European countries with welfare state histories and pervasive housing policies that built subsidized housing at the periphery of major and smaller cities, a new trend of displacement of people from the centre of the city to the periphery due to inner-city gentrification is now a political or at least a planning topic (see Frank 2008).
6 The notion of an inland African American suburb in the Los Angeles region recently received satirical treatment in Paul Beatty's novel *The Sellout* (2015) that depicts the exploits of the inhabitants of the 'agrarian ghetto' of Dickens.
7 'The old racism created the ghetto. The Civil Rights Movement opened its gates and a new black middle class emerged. But the new form of symbolic racism emanating from the iconic ghetto hovers, stigmatizing by degrees black people as they navigate the white space' (Anderson 2015).

6 Beyond the Picket Fence: Global Suburbia

1 I am indebted to Steven Logan's fine (forthcoming) book for this insight. Logan discusses the importance of the house and the debates around it in the modernist discourses on urbanism between European and North American suburbanization in the twentieth century. Logan's work also points to a larger and more diversified ecology of suburban form than the one schematically discussed in this present argument that focuses on the British-colonial angle predominantly. Morphological and design considerations as well as moral attributions were meshed with Fordist political economies in these mostly transatlantic conversations among architects and planners from Eastern Europe to California. Interestingly, the generalized and in Europe often preferred 'other' of the single-family house was the large housing estate in a modernist 'settlement'.

7 Suburban Infrastructures

1 I am extending here an argument made in a joint publication with Pierre Filion 2016.
2 Arguments of that kind are often rehearsed and have reached the status of a truism in sub/urbanist debates on mobilities. In the usual transport land use feedback cycle, for example, the assumption is made that 'suburbanization and the growth in car use have mutually reinforced each other' which is, of course a grave concern in an age of climate change (Bertolini 2012: 19).

8 The Urban Political Ecology of Suburbanization

1 This section is partially based on Keil and Macdonald 2016.
2 (http://villes-noires.tumblr.com/post/88074626960/the-uninhabitable-part-one)

References

Addie, J. -P. D. 2014. Flying High (in the Competitive Sky): Conceptualizing the Role of Airports in Global City-Regions through 'Aero-Regionalism'. *Geoforum*, 55(1), 87–99.

Addie, J. -P. D. 2016. Theorizing Suburban Infrastructure: A Framework for Critical and Comparative Analysis. *Transactions of the Institute of British Geographers*, 41, 3: 273–85.

Addie, J. -P. D. 2017. Governing the Networked Metropolis: The Regionalization of Urban Transportation in Southern Ontario, In R. Keil, P. Hamel, J. -A. Boudreau and S. Kipfer, eds. 2017. *Governing Cities Through Regions: Canadian and European Perspectives*. Waterloo: Wilfrid Laurier University Press.

Addie, J. -P. D. and R. Keil. 2015. Real Existing Regionalism: The Region Between Talk, Territory and Technology. *International Journal of Urban and Regional Research* 39, 2: 407–17.

Addie, J. -P. D., R. Keil, K. Olds. 2014. Beyond town and gown: Higher education institutions, territoriality and the mobilization of new urban structures in Canada. *Territory, Politics, Governance*. DOI:10.1080/21622671.2014.924875, 1–24.

Aguilar, A. G., P. M. Ward, C. B Smith Sr. 2003. Globalization, regional development and mega-city expansion in Latin America: Analyzing Mexico City's peri-urban hinterland. *Cities* 20(1) (February): 3–21.

Ahmed-Ullah, N. 2016. Brampton, a.k.a Browntown. In Pitter, J. and J. Lorinc, eds. *Subdivided: City-Building in an Age of Hyper-Diversity*. Toronto: Coach House Books, 242–53.

Ajl, M. 2014. The Hypertrophic City Versus the Planet of Fields. In Brenner, N., ed., *Implosions / Explosions*. Berlin: Jovis, 533–50.

Allahwala, A. and R. Keil. 2012. *Between Ethnoburb and Slums in the Sky: Toronto's Immigration Geography Today*. IBA Hamburg: Jovis Verlag.

Amati, M. and Taylor, L. 2010. From Green Belts to Green Infrastructure. *Planning Practice and Research*, 25,2: 143–55.

REFERENCES

Amin, A. 2013. Telescopic Urbanism and the Poor, *City: Analysis of Urban Trends, Culture, Theory, Policy, Action*, 17,4: 476–92.

Anacker, K. B., ed. 2015. *The New American Suburb: Poverty, Race and the Economic Crisis*. Farnham and Burlington, VT: Ashgate.

Anderson, E. 2015. The White Space. *Sociology of Race and Ethnicity* 1(1) (January). DOI:10.1177/2332649214561306, 10–21.

Angel, Shlomo, Jason Parent, Daniel L. Civco and Alejandro M. Blei 2010. The Persistent Decline in Urban Densities: Global and Historical Evidence of 'Sprawl', *Lincoln Institute of Land Policy Working Paper*.

Angel, Shlomo, Jason Parent and Daniel L. Civco. 2010. The Fragmentation of Urban Footprints: Global Evidence of Sprawl, 1990–2000, *Lincoln Institute of Land Policy Working Paper*.

Angelo, H. and D. Wachsmuth. 2015. Urbanizing Urban Political Ecology: A Critique of Methodological Cityism. *International Journal of Urban and Regional Research*, 39(1) (January): 16–27.

Antonioni, M. 1970. *Zabriskie Point*. Drama Film (United States).

Archer, J., P. J. P. Sandul and K. Solomonson, eds. 2015. *Making Suburbia: New Histories of Everyday America*. Minneapolis: University of Minnesota Press.

Arboleda, M. 2015. In the Nature of the Non-City: Expanded Infrastructure Networks and the Political Ecology of Planetary Urbanisation. *Antipode* 48, 2: 233–51.

Bakker, K. 2010. *Privatizing Water: Governance Failure and the World's Urban Water Crisis*. Ithaca and London: Cornell University Press.

Badger, E. 2014. Urban forms of racial inequality are migrating to the suburbs. *Oregon Live*. 10 December. http://www.oregonlive.com/opinion/index.ssf/2014/12/urban_forms_of_racial_inequali.html.

Bain, A. 2013. *Creative Margins: Cultural Production in Canadian Suburbs*. Toronto: University of Toronto Press.

Balducci, A. 2012. Planning the Post-Metropolis, *disP – The Planning Review*, 48, 4–5.

Banham, R. 1970. *Los Angeles: The Architecture of Four Ecologies*. London: Penguin Press.

Barau, A. S. and A. N. M. Ludin. 2012. Intersection of Landscape, Anthropocene and Fourth Paradigm. *Living Reviews in Landscape Research* 6 (2012): 5–30.

Barraclough, L. R. 2011. *Making the San Fernando Valley: Rural Landscapes, Urban Development and White Privilege*. Athens, GA: University of Georgia Press.

Basten, L. 2016. From Urban Containment to suburban boundary zones? Greenbelts, greenbelt planning and its governance in the Ruhr area of Germany. Paper presented at the workshop Blue-Green Boundaries in a Suburbanizing World, Belo Horizonte, Brazil, 1 March.

Baum-Snow, N. 2007. Did Highways Cause Suburbanization? *Quarterly Journal of Economics* 122 (2): 775–805. DOI:10.1162/qjec.122.2.775.
Beatty, P. 2015. *The Sellout*. New York: Picador.
Beauregard, R. 2006. *When America Became Suburban*. Minneapolis: University of Minnesota Press.
Beiser, V. 2015. The Deadly Global War for Sand. *Wired*. 26 March. Available at https://www.wired.com/2015/03/illegal-sand-mining/.
Bernt, M. 2009. Partnership for Demolition: The Governance of Urban Renewal in East Germany's Shrinking Cities. *International Journal of Urban and Regional Research* 33, 3: 754–69.
Bernt, M. 2015. The Limits of Shrinkage: Conceptual Pitfalls and Alternatives in the Discussion of Urban Population Loss. *International Journal of Urban and Regional Research* 40(2) (March): 441–50.
Bertolini, L. 2012. Integrating Mobility and Urban Development Agendas: a Manifesto. *disP – The Planning Review* 48, 1: 16–26.
Bloch, R. 1994. *The Metropolis Inverted: the Rise and Shift to the Periphery and the Remaking of the Contemporary City*. Ph.D. dissertation. Los Angeles, CA: UCLA.
Bloch, R. 2015. Africa's New Suburbs. In P. Hamel and R. Keil, eds. *Suburban Governance: A Global View*. Toronto: University of Toronto Press, 253–77.
Bloch, R., N. Papachristodoulou and D. Brown. 2013. Suburbs at Risk. In R. Keil. ed. *Suburban Constellations: Governance, Land and Infrastructure in the 21st Century*. Berlin: Jovis Verlag: 95–101.
Bormann, O. M. Koch and M. Schumacher. 2015. Die Widerfindung der Stadt in der Suburbia. In J. Jessen and F. Roost, eds. *Refitting Suburbia: Erneuerung der Stadt des 20. Jahrhunderts in Deutschland und den USA*. Berlin: Jovis, 113–29.
Boudreau, J. -A. 2016. *Global Urban Politics*. Cambridge: Polity.
Boustan, L. P. and R. A. Margo. 2013. A Silver Lining to White Flight? White Suburbanization and African–American Homeownership, 1940–1980. *Journal of Urban Economics* 78 (November): 71–80.
Bouton, S., S. M. Knupfer, I. Mihov and S. Swartz. 2015. Urban mobility at a tipping point, McKinsey Corporation; available at http://www.mckinsey.com/insights/sustainability/urban_mobility_at_a_tipping_point.
Bouzarovski, S. 2016. *Retrofitting the City: Residential Flexibility, Resilience and the Built Environment*. London and New York: I. B. Tauris.
Brantz, Dorothee B., S. Disko and G. Wagner-Kyora, eds. 2012. *Thick Space: Approaches to Metropolitanism*. Bielefeld: transcript.
Brenner, N. ed. 2014. *Implosions/Explosions: Towards a Study of Planetary Urbanization*. Berlin: Jovis Verlag.

Brenner, N. and C. Schmid 2015. Towards a new epistemology of the urban? *CITY*. 19, 2–3: 151–182, http://dx.doi.org/10.1080/13604813.2015.1014712

Bréville, B. 2016. Mobile homes can't move on. *Le Monde diplomatique*, March: 12–13.

Brugmann, J. 2009. *Welcome to the Urban Revolution*. Toronto: Viking Canada.

Bunnell, T., A. Maringanti. 2010. Practising Urban and Regional Research Beyond Metrocentricity. *International Journal of Urban and Regional Research*. 34: 415–20.

Bunting, T., P. Filion and H. Priston. 2002. Density Gradients in Canadian Metropolitan Regions, 1971–96: Differential Patterns of Central Area and Suburban Growth and Change. *Urban Studies* 39(13): 2531–52.

Burdett, R. and D. Sudjic, eds. 2007. *The Endless City*. London: Phaidon.

Burdett, R. and D. Sudjic. 2011. *Living in the Endless City*. London: Phaidon.

Burns, R. 2014. Sprawled Out in Atlanta, *PoliticoMagazine*, 8 May; available at http://www.politico.com/magazine/story/2014/05/sprawled-out-in-atlanta-106500.

Buxton, M. 2011. Greenbelt and peri-urban resilience to fundamental change. *Global Greenbelts conference, local solutions to global challenges; Closing panel: five big ideas for greenbelts*, 24 March 2011 Toronto.

Caldeira, T. 2013. São Paulo: The City and Its Protest. *openDemocracy*. 11 July. https://www.opendemocracy.net/opensecurity/teresa-caldeira/s%C3%A3o-paulo-city-and-its-protest.

Caldeira, T. 2016. Peripheral Urbanization: Autoconstruction, Transversal Logics and Politics in the Cities of the Global South. *EPD Society and Space*. 35, 1: 3–20.

Campbell, Duncan 2005. *Maputo: an African 'success story' but 80 percent still live in slums*. Available at http://www.theguardian.com/world/2005/feb/02/hearafrica05.development4?CMP=twt_gu.

Capps, Kriston 2015. White People Aren't Driving Growth in the Suburbs: The decline of white suburbia has already begun. The Atlantic CityLab, 29 July; available at http://www.citylab.com/housing/2015/07/white-people-arent-driving-growth-in-the-suburbs/399659/?utm_source=SFFB.

Carpio, G., C. Irazabal and L. Pulido. 2011. Right to the Suburb? Rethinking Lefebvre and Immigrant Activism. *Journal of Urban Affairs* 33, 2: 185–208.

Carter-Whitney, M. 2010. Ontario's greenbelt in an international context. Friends of the greenbelt foundation[online]. Available from: http://www.cielap.org/pdf/GreenbeltInternationalContext2010.pdf.

Castells, M. 1976. The Wild City. *Kapitalistate*, 4–5: 2–30.

Castells, M. 1977. *The Urban Question. A Marxist Approach* (A. Sheridan, translator). London: Edward Arnold. [Original publication in French, 1972].

Cervero, R. 2013. Linking Urban Transport and Land Use in Developing Countries. *Journal of Transport and Land Use* 6(1): 7–24.

Cervero, R. and A. Golub. 2007. Informal Transport: A Global Perspective. *Transport Policy* 14(6): 445–57.

Cervero, R. and J. Day. 2010. Effects of Residential Relocation on Household and Commuting Expenditures in Shanghai, China. *International Journal of Urban and Regional Research* 34(4): 762–88.

Chakrabarti, V. 2013. *A Country of Cities: A Manifesto for an Urban America.* New York: Metropolis Books.

Channel News Asia 2016. Zika fuelled by rapid urbanization, poor conditions in Latam's slums – experts, http://www.channelnewsasia.com/news/world/zika-fuelled-by-rapid-urb/2531488.html.

Charmes, E. 2005. Préface de Jean Rémy. *La Vie périurbaine face à la menace des gated communities.* Paris: L'Harmattan.

Charmes, E and R. Keil. 2015. The Politics of Post-Suburban Densification in Canada and France. *International Journal of Urban and Regional Research.* 581–602. DOI:10.1111/1468-2427.12194.

Cheng, W. 2014. *The Changs Next Door to the Díazes: Remapping Race in Suburban California.* Minneapolis: University of Minnesota Press.

CityMayors Statistics 2007. *The Largest Cities in the World by Land Area, Population and Density.* Available at http://www.citymayors.com/statistics/largest-cities-density-125.html.

Clapson, M. and R. Hutchison. 2010. *Suburbanization in Global Society.* Emerald Group Publishing.

Cohen, D. A. 2014. Seize the Hamptons. *Jacobin Magazine* 15–16: 151–59.

Coker, A. 2016. *Negotiating Informal Housing in Metro Manila: Forging Communities through Participation.* Jyväskylä: University of Jyväskylä.

Cowen, D. and N. Lewis. 2016. Anti-blackness and urban geopolitical economy: Reflections on Ferguson and the suburbanization of the 'internal colony', *Society & Space*; available at https://societyandspace.com/material/commentaries/deborah-cowen-and-nemoy-lewis-anti-blackness-and-urban-geopolitical-economy-reflections-on-ferguson-and-the-suburbanization-of-the-internal-colony/.

Cox, W. 2012. Density is Not the Issue: The Urban Scaling Research. Available at http://www.newgeography.com/content/002987-density-not-issue-the-urban-scaling-research.

Crawford, M. 2013. Little Boxes: High Tech and the Silicon Valley, *Room One Thousand*, Issue 1; http://www.roomonethousand.com/index#/little-boxes-high-tech-and-the-silicon-valley/.

Crawford, M. 2015. Afterword. In Archer, J., P. J. P. Sandul and K. Solomonson, eds. *Making Suburbia: New Histories of Everyday America.* Minneapolis: University of Minnesota Press, 381–7.

Crutzen, P. J. 2005. Human Impact on Climate has made this the 'Anthropocene Age'. *New Perspectives Quarterly* 22,2: 14–16.

Cutler, K. M. 2014. How Burrowing Owls Lead To Vomiting Anarchists (Or SF's Housing Crisis Explained); *TechCrunch* 14 April; available at http://techcrunch.com/2014/04/14/sf-housing/?ncid=twittersocialshare.

Davis, M. 1990. *City of Quartz: Excavating the Future in Los Angeles*. London: Verso.

Davis, M. 1994. The Suburban Nightmare: While older suburbs experience many problems of the inner city, 'edge cities' now offer a new escape, *Los Angeles Times*, 23 October, available at http://articles.latimes.com/1994-10-23/opinion/op-53893_1_edge-cities.

Davis, M. 1995. Los Angeles after the storm: The dialectic of ordinary disaster. *Antipode* 27, 3: 221–41.

Davis, M. 2006. *Planet of Slums*. London. New York: Verso.

Davis, M. 2011. Joblessness adds heat to Imperial Valley protests, *Los Angeles Times*, 8 November, available at http://articles.latimes.com/2011/nov/08/opinion/la-oe-davis-elcentro-20111108.

Dear, M. and N. Dahmann. 2011. Urban Politics and the Los Angeles School of Urbanism. In D. Judd and D. Simpson, eds. *The City, Revisited: Urban Theory from Chicago, Los Angeles, New York*. Minneapolis: University of Minnesota Press, 65–78.

De Jong, J. 2014. *New Suburbanisms*. London: Routledge.

De Meyer, D., K. Versluys, K. Borret, B Eeckhout, S. Jacobs and B. Keunen (Ghent Urban Studies Team). 1999. *The Urban Condition: Space, Community and Self in the Contemporary Metropolis*. Rotterdam: 010 Publishers.

Demos, T. J. 2015. IV. Capitalocene Violence, http://blog.fotomuseum.ch/2015/06/iv-capitalocene-violence/.

Desmond, M. 2017. How Homeownership Became the Engine of American Inequality, *The New York Times Magazine*, May 9, https://www.nytimes.com/2017/05/09/magazine/how-homeownership-became-the-engine-of-american-inequality.html.

Dikeç, M. 2007. *Badlands of the Republic: Space, Politics and Urban Policy*. Malden, MA; Oxford: Blackwell Publishers.

Dikeç, M. and E. Swyngedouw. 2017. Theorizing the Politicizing City. *International Journal of Urban and Regional Research*. [online early]

Dillon, L. 2017. California won't meet its climate change goals without a lot more housing density in its cities, *Los Angeles Times*, 6 March; available at http://www.latimes.com/politics/la-pol-ca-housing-climate-change-goals-20170306-story.html.

Dodson, J. 2007. *The Australian Suburb as a Socio-Technical Process: Towards a Research Agenda*. Urban Research Program, ENV, Griffith University.

Dodson, J. 2014. 'Suburbia under an Energy Transition: A Socio-technical Perspective'. *Urban Studies* 51(7) (May): 1487–505.

Dodson, J. and N. Sipe. 2009. A Suburban Crisis? Housing, Credit, Energy and Transport. *Journal of Australian Political Economy* 64: 199–210.

Doshi, V. 2016. Gurgaon: what life is like in the Indian city built by private companies. *Guardian*, 4 July; available at https://www.theguardian.com/sustainable-business/2016/jul/04/gurgaon-life-city-built-private-companies-india-intel-google?CMP=twt_gu.

Dossick C. S., L. Dunn, I. Fishburn, N. Gualy, K. R. Merlino and J. Twill. 2012. *The Conflicted City Hypergrowth, Urban Renewal and Mass Urbanization in Istanbul*. Runstad Center for Real Estate Studies, University of Washington. http://realestate.washington.edu/wp-content/uploads/2013/03/ConflictedCity.pdf.

Dreier, P. and T. Swanstrom. 2014. Suburban ghettos like Ferguson are ticking time bombs. *The Washington Post*, 21 August; available at https://www.washingtonpost.com/posteverything/wp/2014/08/21/suburban-ghettos-like-ferguson-are-ticking-time-bombs/.

Drummond, L. and D. Labbé. 2013. We're a Long Way from Levittown, Dorothy: Everyday Suburbanism as a Global Way of Life. In R. Keil, ed. *Suburban Constellations: Governance, Land and Infrastructure in the 21st Century*. Berlin: Jovis Verlag, 46–51.

Dunham-Jones, E. 2015. Umgestaltung von Suburbia: Ein amerikanischer oder ein globaler Trend. In J. Jessen and F. Roost, eds. *Refitting Suburbia: Erneuerung der Stadt des 20. Jahrhunderts in Deutschland und den USA*. Berlin: Jovis, 95–112.

Dunham-Jones, E. and J. Williamson 2009. *Retrofitting Suburbia: Urban Design Solutions for Redesigning Suburbs*. Hoboken: John Wiley and Sons.

Dunham-Jones, E. and J. Williamson. 2011. *Retrofitting Suburbia: Urban Design Solutions for Redesigning Suburbs*. Updated edn. Hoboken: Wiley & Sons.

Easterling, K. 2014. *Extrastatecraft: The Power of Infrastructure Space*. London and New York: Verso.

Economist 2008. An age of transformation. *The Economist*, 29 May, available at www.economist.com.

Economist 2016. Where do Canada's immigrants come from? *The Daily Chart*, 28 October; available at http://www.economist.com/blogs/graphicdetail/2016/10/daily-chart-18?fsrc=scn/tw/te/bl/ed/.

Edensor, T. and M. Jayne, eds. 2012. *Urban Theory Beyond the West: A World of Cities*. London and New York: Routledge.

Ehrenhalt, A. 2012. *The Great Inversion and the Future of the American City*. New York: Knopf.

Ekers, M., P. Hamel and R. Keil. 2012. Governing Suburbia: Modalities and Mechanisms of Suburban Governance. *Regional Studies* 46(3): 405–22.

Enright, T. 2016. *The Making of Grand Paris: Metropolitan Urbanism in the Twenty-First Century*. Cambridge, Massachusetts: The MIT Press.

Epstein, Renaud 2013. Le 'problème des banlieues' après la désillusion de la rénovation, *Métropolitiques*, 18 janvier. http://www.metropolitiques.eu/Le-probleme-des-banlieues-apres-la.html.

Farias, I. and S. Stemmler. 2012. Deconstructing 'Metropolis:' Critical Reflections on a European Concept. In D. Brantz, S. Disko and G. Wagner-Kyora, eds. *Thick Space: Approaches to Metropolitanism*. Bielefeld: Transcript, 49–66.

Farrell T. and S. Kelly. 2015. Ballymun: From high hopes to broken dreams, *Irish Examiner*, 3 August; http://www.irishexaminer.com/ireland/ballymun-from-high-hopes-to-broken-dreams-345936.html.

Fiedler, R. and J-P. Addie. 2008. Canadian Cities on the Edge. *The City Institute at York University: Occasional Paper Series* 1(1): 1–33.

Filion, P. 2013a. The Infrastructure is the Message: Shaping the Suburban Morphology and Lifestyle. In R. Keil. ed. *Suburban Constellations: Governance, Land and Infrastructure in the 21st Century*. Berlin: Jovis, 39–45.

Filion, P. 2013b. Automobiles, Highways and Suburban Dispersion. In R. Keil. ed. *Suburban Constellations: Governance, Land and Infrastructure in the 21st Century*. Berlin: Jovis, 79–84.

Filion, P. 2015. Suburban Inertia: The Entrenchment of Dispersed Suburbanism. Forthcoming in *International Journal of Urban and Regional Research*. 39, 3: 633–40.

Filion, P. and R. Keil. 2016. Contested Infrastructures: Tension, Inequity and Innovation in the Global Suburb. *Urban Policy and Research*. http://dx.doi.org/10.1080/08111146.2016.1187122.

Filion, P. and N. Pulver, eds. Forthcoming. *Global Suburban Infrastructure: Social Restructuring, Governance and Equity*. Toronto: University of Toronto Press.

Filion, P., K. McSpurren and B. Appleby. 2006. 'Wasted Density? The Impact of Toronto's Residential-density-distribution Policies on Public-transit Use and Walking'. *Environment and Planning A* 38(7): 1367–92.

Filler, M. 2016. Living Happy Ever After. *The New York Review of Books*. 21 April; http://www.nybooks.com/articles/2016/04/21/houses-living-happily-ever-after/.

Fishman, R. 1987. *Bourgeois Utopias: The Rise and Fall of Suburbia*. New York: Basic Books.

Fleischer, F. 2010. *Suburban Beijing: Housing and Consumption in Contemporary China*. Minneapolis: University of Minnesota Press.

Flew, T., M. Gibson, C. Collis and E. Felton. 2012. Creative Suburbia: Cultural Research and Suburban Geographies. *International Journal of Cultural Studies* 15(3): 199–203.

Florida, R. 2002. *The Rise of the Creative Class: And How It's Transforming Work, Leisure, Community and Everyday Life*. New York: Basic Books.

Florida, R. 2016. The Role of Cities in Preventing Crisis: A conversation with the scholar Josef Konvitz. Available at http://www.citylab.com/politics/2016/05/the-role-of-cities-in-preventing-crisis/483746/.

Fogelson, R. 2005. *Bourgeois Nightmares: Suburbia, 1870–1930*. New Haven and London: Yale University Press.

Fontenot, A. 2015. Notes Toward a History of Non-Planning: On design, the market and the state. *Places Journal* (January) Available at https://placesjournal.org/article/notes-toward-a-history-of-non-planning/#.

Ford, L. R. 1994. *Cities and Buildings: Skyscrapers, Skid Rows and Suburbs*. Baltimore and London: The Johns Hopkins University Press.

Forsyth, A. 2012. Defining Suburbs. *Journal of Planning Literature* 27, 3: 270–81.

Forsyth, A. 2014. Global Suburbia and the Transition Century: Physical Suburbs in the Long Term. *Urban Design International* 19, 4: 259–73.

Fortin, A. The Suburbs and the Bungalow Heritage in the Making. The Encyclopedia of French Cultural Heritage in North America. http://www.ameriquefrancaise.org/en/article-612/The_suburbs_and_the_bungalow_heritage_in_the_making.html.

Frank, C. 2008. Vertreibung aus dem reichen Herz der Städte, *Süddeutsche Zeitung*, 2 July; available at www.sueddeutsche.de.

Freund, D. M. P. 2016. We can't forget how racist institutions shaped homeownership in America, *The Washington Post*, 28 April, available at https://www.washingtonpost.com/news/wonk/wp/2016/04/28/we-cant-forget-how-racist-institutions-shaped-homeownership-in-america/.

Frey, W. 2014. The suburbs: not just for white people anymore. *The New Republic* (available at http://www.newrepublic.com/article/120372/white-suburbs-are-more-and-more-thing-past.

Friedman, A. 2013. *Covert Capital: Landscapes of Denial and the Making of U. S. Empire in the Suburbs of Northern Virginia*. Berkeley: University of California Press.

Friedmann, J. 2002. *The Prospect of Cities*. Minneapolis: University of Minnesota Press.

Friedmann, J. and J. Miller 1965. The Urban Field. *Journal of the American Institute of Planners* 31, 4: 312–20.

Gallagher, L. 2013. *The End of Suburbia: Where the American Dream is Moving*. New York: Penguin.

Gandy, M. 2014. *The Fabric of Space: Water, Modernity and the Urban Imagination*. Cambridge, MA: MIT Press.

Gans, H. 1967. *The Levittowners: Ways of Life and Politics in a New Suburban Community*. New York: Pantheon Books.

Garreau, J. 1991. *Edge City. Life on the New Frontier*. New York: Anchor Books.

Gee, M. 2017. Spillover: When the City Comes to the Country. *The Globe and Mail*, 4 March: M1–4.

Gibson, C., C. Brennan-Horley, B. Laurenson, N. Riggs, A. Warren, B. Gallan and H. Brown. 2012. 'Cool places, creative places? Community perceptions of cultural vitality in the suburbs'. *International Journal of Cultural Studies* 15(3): 287–302.

Gilbert, L. and F. De Jong. 2015. Entanglements of Periphery and Informality in Mexico City. *International Journal of Urban and Regional Research*. 39, 3: 518–32.

Glaeser, E. 2011. *Triumph of the City: How Our Best Invention Makes Us Richer, Smarter, Greener, Healthier and Happier*. New York: Penguin Press.

Gleeson, B. 2014. What Role for Social Science in the 'Urban Age'? In N. Brenner, ed., *Implosions/Explosions: Towards a Study of Planetary Urbanization*. Berlin: Jovis Verlag, 338–52.

Gordon, I. 2016. Next London Mayor should work with Wider South East to rethink the green belt, LSE Centre for Cities. Available at http://www.centreforcities.org/blog/next-london-mayor-work-wider-south-east-rethink-green-belt/.

Gordon, D. L. A. and M. Janzen. 2013. Suburban Nation? Estimating the Size of Canada's Suburban Population. *Journal of Architectural and Planning Research* 30, 3: 197.

Gottdiener, M. 1977. *Planned Sprawl*. Beverly Hills and London: Sage Publications.

Graham, S. and S. Marvin. 2001. *Splintering Urbanism: Networked Infrastructures, Technological Mobilities and the Urban Condition*. London: Routledge.

Greason, W. D. 2012. *Suburban Erasure: How the Suburbs Ended the Civil Rights Movement in New Jersey*. New Jersey: Fairleigh Dickinson University Press.

Greenfield, A. 2016. Where are the world's newest cities...and why do they all look the same? *The Guardian Cities*, 28 January. http://www.theguardian.com/cities/2016/jan/28/where-world-newest-cities-look-same.

Grunwald, M. 2015. Overpasses. A Love Story. Politico (August); Available at http://www.politico.com/agenda/issue/transportation-august-2015.

Guardian 2015. Facebook offers employees $10,000 to live close to the office, Guardian, 15 December, available at https://www.theguardian.com/technology/2015/dec/18/facebook-offers-employees-10000-to-live-close-to-the-office?utm_source=esp&utm_medium=Email&utm_campaign=GU

+Today+main+NEW+H&utm_term=144848&subid=12447916&CMP= EMCNEWEML661912.

Guerra, E. 2016. Planning for Cars That Drive Themselves: Metropolitan Planning Organizations, Regional Transportation Plans and Autonomous Vehicles. *Journal of Planning Education and Research* 36(2): 210–24.

Guney, M., R. Keil, M. Ucoglu, eds. Forthcoming. *Massive Suburbanization*. Toronto: University of Toronto Press.

Gururani, S. 2013. On Capital's Edge: Gurgaon: India's Millennial City. In R. Keil, ed. *Suburban Constellations: Governance, Land and Infrastructure in the 21st Century*. Berlin: Jovis Verlag, 182–9.

Gururani, S. Forthcoming. Cities in a World of Villages: Tracking the urban through the agrarian. *Urban Geography*.

Gururani, S. and B. Kose. 2015. Shifting Terrain: Questions of Governance in India's Cities and Their Peripheries. In P. Hamel and R. Keil, eds. *Suburban Governance: A Global View*. Toronto: University of Toronto Press, 278–302.

Hae, L. Forthcoming. The 'Construction State' Unbound: Variegated Neoliberal Urbanization and Struggles over Greenbelt Deregulation in the Seoul Metropolitan Region. *International Journal of Urban and Regional Research*.

Haila, A. 2015. *Urban Land Rent: Singapore as a Property State*. Oxford: Wiley Blackwell.

Haldeman, P. 2015. The Building of Bigger Mansions Leaves Los Angeles in the Dust. *The New York Times International Weekly*, Weekend, 31 January–1 February, p. 6.

Hamel, P. and R. Keil, eds. 2015. *Suburban Governance: A Global View*. Toronto: University of Toronto Press.

Hanlon, B., J. R. Short and T. J. Vicino. 2010. *Cities and Suburbs: New Metropolitan Realities in the US*. New York: Routledge.

Harlander, T. 2011. Die 'Modernität' der Boomjahre: Flächensanierung und Großsiedlungsbau, *ARCH+*, 44, 203, June: 18.

Harris, C. 2004. How Did Colonialism Dispossess? Comments from an Edge of Empire, *Annals of the Association of American Geographers*, 94(1), 165–82.

Harris, R. 2004. *Creeping Conformity: How Canada Became Suburban, 1900–1960*. Toronto: University of Toronto Press.

Harris, R. 1996. *Unplanned Suburbs: Toronto's American Tragedy, 1900–1950*. Baltimore and London: The Johns Hopkins University Press.

Harris, R. 2010. Meaningful Types in a World of Suburbs. In M. Clapson and R. Hutchinson, eds. *Suburbanization in Global Society*. Bingley, UK: Emerald, 15–50.

Harris, R. 2013. How Land Markets Make and Change Suburbs. In R. Keil, ed. *Suburban Constellations: Governance, Land and Infrastructure in the 21st Century*. Berlin: Jovis.

Harris, R. 2014. Using Toronto to Explore Three Suburban Stereotypes and Vice Versa. *Environment and Planning A* 46(1): 30–49.

Harris, R. and P. J. Larkham. 1999. Suburban Foundation, Form and Function. In Harris, R. and P. J. Larkham, eds. *Changing Suburbs. Foundation, Form and Function.* New York: E. and F. N. Spon.

Harris, R. and U. Lehrer, eds. 2018. *The Suburban Land Question: A Global Survey.* Toronto: University of Toronto Press.

Harris, R. and C. Vorms, eds. 2017. *What's in a Name? Talking about 'Suburbs'.* Toronto: University of Toronto Press.

Harvey, D. 1982. *The Limits to Capital.* Oxford: Blackwell.

Harvey, D. 1996. *Justice, Nature and the Geography of Difference.* Oxford: Blackwell.

Harvey, D. 2007. The Right to the City. *New Left Review*, 53 (Sept/Oct), 23–40.

Harvey, D. 2013. New David Harvey Interview on Class Struggle in Urban Spaces. Critical-Theory, 7 August 2013. http://www.critical-theory.com/david-harvey-interview-class-struggle-urban-spaces/.

Hayden, D. 2003. *Building Suburbia: Green Fields and Urban Growth, 1820–2000.* New York: Pantheon Books.

Heeg, S. 2017. Governing the Built Environment in European Metropolitan Regions: Financialization, Responsibilization and Urban Competition. In R. Keil, P. Hamel, J. -A. Boudreau and S. Kipfer, eds. 2017. *Governing Cities Through Regions: Canadian and European Perspectives.* Waterloo: Wilfrid Laurier University Press: 65–82.

Heiman, R. 2015. *Driving After Class: Anxious Times in an American Suburb.* Oakland, California: University of California Press.

Heinrichs, D. and H. Nuissl. 2015. Suburbanization in Latin America: Towards New Authoritarian Modes of Governance at the Urban Margin. In P. Hamel and R. Keil, eds. *Suburban Governance: A Global View.* Toronto: University of Toronto Press, 216–38.

Hertel, S., R. Keil and M. Collens. 2015. *Switching Tracks: Towards transit equity in the Greater Toronto and Hamilton Area.* Toronto: The City Institute at York University.

Hertel, S., R. Keil and M. Collens. 2016. *Next Stop Equity: Routes to fairer transit access in the Greater Toronto and Hamilton Area.* Toronto: The City Institute at York University.

Herzog, L. 2015. *Global Suburbs: Urban Sprawl from the Rio Grande to Rio de Janeiro.* London: Routledge.

Hesse, M. 2013. Cities and Flows: Re-Asserting a Relationship as Fundamental as it is Delicate. *Journal of Transport Geography* 29: 33–42.

Hesse, M. 2015. Suburbs: the next slum? Explorations into the contested terrain of social construction and political discourse. *Articulo – revue de sci-*

ences humaines [En ligne], mis en ligne le 20 décembre 2015. URL: http://articulo.revues.org/1552

Heynen, N. C. 2014. Urban Political Ecology 1: The Urban Century. *Progress in Human Geography* 38, 4: 598–604.

Heynen, N., M. Kaika and E. Swyngedouw. 2006. Urban political ecology: politicizing the production of urban natures. In N. C. Heynen, M. Kaika and E. Swyngedouw, eds. *In the Nature of Cities: Urban Political Ecology and the Politics of Urban Metabolism*. Abingdon: Routledge, 1–20.

Hirt, S. 2012. *Iron Curtains: Gates, Suburbs and Privatization of Space in the Post-Socialist City*. Oxford: Wiley-Blackwell.

Hirt, S. and A. Kovachev. 2015. Suburbia in Three Acts: The East European Story. In P. Hamel and R. Keil, eds. *Suburban Governance: A Global View*. Toronto: University of Toronto Press, 177–97.

Hoch, C. 1985. Municipal Contracting in California: Privatizing with Class. *Urban Affairs Review* 20,3: 303–23.

Hodson, M. and S. Marvin. 2010. Urbanism in the Anthropocene: Ecological Urbanism or Premium Ecological Enclaves? *City* 14,3: 298–313.

Hodson, M. and S. Marvin, eds. 2016. *Retrofitting Cities: Priorities, Governance and Experimentation*. London: Routledge.

Hochschild, A. R. 2016a. *Strangers in their Own Land: Anger and Mourning on the American Right*. New York and London: The New Press.

Hochschild, A. R. 2016b. How the 'Great Paradox' of American politics holds the secret to Trump's success, *Guardian*, 7 September; Available at https://www.theguardian.com/us-news/2016/sep/07/how-great-paradox-american-politics-holds-secret-trumps-success.

Horkheimer, M. and T. W. Adorno. 1972. *Dialectic of Enlightenment*. New York: Continuum.

Hudalah, D., H. Winarso and J. Woltjer. 2016. Gentrifying the Peri-Urban: Land Use Conflicts and Institutional Dynamics at the Frontier of an Indonesian Metropolis. *Urban Studies* 53, 3: 593–608.

Hughes, T. 1987. The Evolution of Large Technological Systems. In W. Bijker, T. Hughes and T. Pinch, eds. *The Social Construction of Technological Systems*, Cambridge, MA and London, UK, 50–82.

Hulchanski, D. 2010. *The Three Cities Within Toronto. Income Polarization Among Toronto's Neighbourhoods, 1970–2005*. Toronto: University of Toronto.

Hume, C. 2015. It is time to transform the suburbs. *Toronto Star*, 13 February. Available at http://www.thestar.com/news/gta/2015/02/13/its-time-to-transform-the-suburbs-hume.html.

Humphreys, C. 2016. Cities are not as big a deal as you think. *Undark*. 6 June. https://undark.org/article/cities-suburbs-urban-social-environmental/.

Huq, R. 2013. *Making Sense of Suburbia through Popular Culture*. London: Bloomsbury.
Jacobs, J. 1969. *The Economy of Cities*. London: Jonathan Cape.
Jacobs, J. 1996. *Edge of Empire: Postcolonialism and the City*. London: Routledge.
Jauhiainen, J. S. 2013. Suburbs. In P. Clark, ed. *The Oxford Handbook of Cities in World History*. Oxford: Oxford University Press, 791–809.
Jeffries, S. 2014. Mordor, he wrote: how the Black Country inspired Tolkien's badlands. *Guardian*. 19 September. Available at http://www.theguardian.com/books/2014/sep/19/how-the-west-midlands-black-country-inspired-tolkien-lord-of-the-rings.
Jessen, J. and F. Roost, eds. 2015a. *Refitting Suburbia: Erneuerung der Stadt des 20. Jahrhunderts in Deutschland und den USA*. Berlin: Jovis.
Jessen, J. and F. Roost. 2015b. Editorial: Refitting Suburbia – Umbau der Siedlungsstrukturen des 20. Jahrhunderts, in *Refitting Suburbia: Erneuerung der Stadt des 20. Jahrhunderts in Deutschland und den USA*. Berlin: Jovis, 7–22.
Johnson, K. S. 2014. Black Suburbanization: American Dream or the New Banlieue?, *The CitiesPapers*, http://citiespapers.ssrc.org/black-suburbanization-american-dream-or-the-new-banlieue/
Johnson, L. C. 2013. Desire, Dryness and Decadence. In R. Keil, ed. *Suburban Constellations: Governance, Land and Infrastructure in the 21st Century*. Berlin: Jovis Verlag, 190–4.
Johnson, L. C. 2015. Governing Suburban Australia. In P. Hamel and R. Keil, eds. *Suburban Governance: A Global View*. Toronto: University of Toronto Press, 110–29.
Jonas, A. E. G., S. Pincetl and J. Sullivan. 2013. Endangered Neoliberal Suburbanism? The Use of the Federal Endangered Species Act as a Growth Management Tool in Southern California. *Urban Studies* 50(11): 2311–31.
Judd, D. R. and D. Simpson, eds. 2011. *The City Revisited: Urban Theory from Chicago, Los Angeles, New York*. Minneapolis: University of Minnesota Press.
Kabisch, S. and D. Rink. 2015. Governing Shrinkage of Large Housing Estates at the Fringe. In P. Hamel and R. Keil, eds. *Suburban Governance: A Global View*. Toronto: University of Toronto Press, 198–215.
Kaika, M. 2015. The Uncanny Materialities of the Everyday: Domesticated Nature as the Invisible 'Other'. In S. Graham and C. McFarlane, eds. *Infrastructural Lives: Urban Infrastructure in Context*. New York and Abingdon: Routledge, Earthscan, 9–52.
Kaika, M. and L. Karaliotas. 2016. The Spatialization of Democratic Politics: Insights from Indignant Squares. *European Urban and Regional Studies* 23(4) (October): 556–70.

Kaika, M. and E. Swyngedouw. 2000. Fetishizing the Modern City: The Phantasmagoria of Urban Technological Networks. *International Journal of Urban and Regional Research*. 4, 1:120–38.

Katz, B. and J. Bradley. 2013. *The Metropolitan Revolution: How Cities and Metros Are Fixing Our Broken Politics and Fragile Economy*. Brookings.

Keil, R. 1998. *Los Angeles: Globalization, Urbanization and Social Struggles*. Chichester: Wiley.

Keil, R. 2003. Progress Report: Urban Political Ecology. *Urban Geography* 24, 8: 723–38.

Keil, R. 2007. Empire and the Global City: Perspectives of Urbanism after 9/11. *Studies in Political Economy* 79: 167–92.

Keil, R. ed. 2013. *Suburban Constellations: Governance, Land and Infrastructure in the 21st Century*. Berlin: Jovis Verlag.

Keil, R. and J. P. Addie. 2015. 'It's Not Going to be Suburban, It's Going to be All Urban': *Assembling Postsuburbia in the Toronto and Chicago Regions*. (September) 39, 5: 892–911.

Keil, R. and P. Hamel. 2015. Conclusion: Suburban Governance: Convergent and Divergent Dynamics. In P. Hamel and R. Keil, eds. *Suburban Governance: A Global View*. Toronto: University of Toronto Press, 349–58.

Keil, R., P. Hamel, J.-A. Boudreau and S. Kipfer, eds. 2017. *Governing Cities Through Regions: Canadian and European Perspectives*. Waterloo: Wilfrid Laurier University Press.

Keil, R., P. Hamel, E. Chou and K. Williams. 2015. Modalities of Suburban Governance in Canada. In P. Hamel and R. Keil, eds. *Suburban Governance: A Global View*. Toronto: University of Toronto Press.

Keil, R. and S. Macdonald. 2016. Rethinking Urban Political Ecology from the Outside in: Greenbelts and Boundaries in the Post-Suburban City. *Local Environment*, DOI:10.1080/13549839.2016.1145642.

Keil, R. and K. Ronneberger. 1994. Going up the Country: Internationalization and Urbanization on Frankfurt's Northern Fringe. *Environment and Planning D: Society and Space* 12(2): 137–66.

Keil, R. and M. Whitehead. 2012. Cities and the Politics of Sustainability. In K. Mossberger, S. E. Clarke and P. John, eds. *The Oxford Handbook of Urban Politics*. New York: Oxford University Press, 520–3.

Keil, R. and D. Young. 2009. Fringe Explosions: Risk and Vulnerability in Canada's New In-Between Urban Landscape, *The Canadian Geographer* 53, 3: 374–86.

Keil, R. and D. Young. 2014. Locating the Urban In-Between: Tracking the Urban Politics of Infrastructure in Toronto. *International Journal of Urban and Regional Research* 38(5): 1589–608.

Kelly, B. M. 2009. Introduction. In D. Rubey, ed. *Redefining Suburban Studies: Searching for New Paradigms*. Hofstra University: The National Center for Suburban Studies, 1–5.

King, A. D. 2004. *Spaces of Global Cultures: Architecture, Urbanism, Identity*. London; New York: Routledge.

Kipfer, S. n. d. *Demolition and Counterrevolution: La Rénovation Urbaine in Greater Paris*. Unpublished. Toronto: York University.

Kipfer, S., P. Saberi, T. Wieditz. 2013. Henri Lefebvre: Debates and Controversies. *Progress in Human Geography* 37, 1: 115–34.

Kling, R., S. Olin and M. Poster. 1995. The Emergence of Postsuburbia: An Introduction. In R. Kling, S. Olin and M. Poster, eds. *Postsuburban California*. Berkeley: University of California Press, 1–30.

Kneebone, E. and A. Berube. 2014. *Confronting Suburban Poverty in America*. Brookings Institution Press.

Kneebone, E. and C. A. Nadeau. 2015. 'The Resurgence of Concentrated Poverty in America: Metropolitan Trends in the 2000s'. In K. B. Anacker, ed. *The New American Suburb: Poverty, Race and the Economic Crisis*, Farnham. Surrey: Ashgate, 15–38.

Knox, P. 2008. *Metroburbia, USA*. New Jersey: Rutgers University Press.

Kohn, M. 2016. *The Death and Life of the Urban Commonwealth*. Oxford: OUP.

Kooy, M. and K. Bakker. 2008. Splintered Networks: The Colonial and Contemporary Waters of Jakarta. *Geoforum* 39: 1843–58.

Kotkin, J. 2013. How Can We Be So Dense? Anti-Sprawl Policies Threaten America's Future. *Forbes Magazine*. Available at http://www.forbes.com/sites/joelkotkin/2013/08/08/how-can-we-be-so-dense-anti-sprawl-policies-threaten-americas-future/.

Kruse, K. M. and T. J. Sugrue. 2006. *The New Suburban History*. Chicago: The University of Chicago Press.

Kunstler, J. H. 1993. *The Geography of Nowhere: The Rise And Decline of America's Man-Made Landscape*. New York: Free Press.

Lang, R., J. LeFurgy and A. C. Nelson. 2006. The Six Suburban Eras of the United States. *Opolis* 2(1): 65–72.

Lang, R. E. and J. B. LeFurgy. 2007. *Boomburbs: The Rise of America's Accidental Cities*. Washington, D.C.: Brookings Institution Press.

Larco, N. 2010. Suburbia Shifted: Overlooked Trends and Opportunities in Suburban Multifamily Housing. *Journal of Architectural and Planning Research* 27, 1: 70–87.

Lawhon, M., H. Ernstson and J. Silver. 2014. 'Provincializing Urban Political Ecology: Towards Situated UPE Through African Urbanism'. *Antipode* 46, 2: 497–516. DOI:10.1111/anti.12051.

Lawton, P., E. Murphy and D. Redmond. 2013. Residential Preferences of the 'Creative Class'? *Cities*. 31: 47–56.

Lazarus, E. D. 2014. Land Grabbing as a Driver of Environmental Change. *Area* 46(1): 74–82.

Lefebvre, H. 1989. Quand la ville se perd dans une métamorphose planétaire, in *Le Monde diplomatique*, May; republished in Manière de voir 114, *Le Monde diplomatique*, December 2010/January 2011, 20–3.

Lefebvre, H. 1991. *The Production of Space*. Oxford: Blackwell.

Lefebvre, H., [ed. by E. Kofman and E. Lebas]. 1996. *Writings on Cities*. Cambridge, MA: Blackwell Publishers.

Lefebvre, H. 2003. *The Urban Revolution*. Minneapolis: University of Minnesota Press.

Lefebvre, H. 2014. Dissolving City, Planetary Metamorphosis. *Environment and Planning D: Society and Space*, 32, 2035.

Lefebvre, H. and K. Ross. 2015. Finding a Larger Theory of the City: Henri Lefebvre und die Situationistische Internationale. *Dérive* 58: 45–50.

Le Galès, P. 2002. *European Cities: Social Conflicts and Governance*. New York: Oxford University Press.

Lehrer, U. 1994 The Image of the Periphery: The Architecture of FlexSpace, *Environment and Planning D: Society and Space*, 12 (2), 187–205.

Lehrer, U., R. Harris and R. Bloch. 2015. La question du territoire suburbain, *Pole Sud: Revue de science de l'Europe méridionale*, special issue Sous le foncier, la politique, 42, 1: 63–85.

Leinberger, C. 2008. The Next Slum? *The Atlantic*. March. https://www.theatlantic.com/magazine/archive/2008/03/the-next-slum/306653/.

Lewinnek, E. 2014. *The Working Man's Reward*. Oxford: Oxford University Press.

Lewis, R., ed. 2004a. *Manufacturing Suburbs: Building Work and Home on the Metropolitan Fringe*. Philadelphia: Temple University Press.

Lewis, R. 2004b. Introduction. In R. D. Lewis, ed. 2004a. *Manufacturing Suburbs: Building Work and Home on the Metropolitan Fringe*. Philadelphia: Temple University Press.

Lewyn, M. 2016. Two Types of Black Suburbanization, *Planetizen*, 3 January; available at: http://www.planetizen.com/node/82983/two-types-black-suburbanization.

Ley, D. and N. Lynch. 2012. Divisions and Disparities in Lotus-Land: Socio-Spatial Income Polarization in Greater Vancouver, 1970–2005. Research Paper 223, Cities Centre, University of Toronto.

Li, W. 1998. Anatomy of a New Ethnic Settlement: The Chinese Ethnoburb in Los Angeles. *Urban Studies* 35(3): 479–501.

Li, W. 2009. *Ethnoburb: The New Ethnic Community in Urban America*. Honolulu: University of Hawai'i Press.

Logan, J. R. 2014. Separate and Unequal in Suburbia. Census Brief prepared for Project US2010. http://www.s4.brown.edu/us2010.

Logan, J. and H. Molotch. 1987. *Urban Fortunes: The Political Economy of Place*. Berkeley: University of California Press.

Logan, S. Forthcoming. *In the Suburbs of History: Modernist Visions of the Urban Periphery*. Toronto: University of Toronto Press.

Lutz, A. 2017. The American suburbs as we know them are dying, *Business Insider*, 5 March; http://www.businessinsider.com/death-of-suburbia-series-overview-2017-3.

Mabin, A. 2013. Suburbanisms in Africa? In R. Keil, ed. *Suburban Constellations: Governance, Land and Infrastructure in the 21st Century*. Berlin: Jovis Verlag, 154–60.

Mabin, A., S. Butcher and R. Bloch. 2013. Peripheries, Suburbanisms and Change in Sub-Saharan African Cities. *Social Dynamics: A Journal of African Studies* 39(2). DOI:10.1080/02533952.2013.796124, 167–90.

Macdonald, S. and L. Lynch. Forthcoming. 'Greenfrastructure': The Ontario Greenbelt as Urban Boundary. In P. Filion and N. Pulver, eds. Forthcoming. *Global Suburban Infrastructure: Social Restructuring, Governance and Equity*. Toronto: University of Toronto Press.

Mace, A., N. A. Phelps and R. Rodieri. 2017. City of villages? Stasis and change in London's suburbs. In N. A. Phelps, ed. *Old Europe, New Suburbanization? Governance, Land and Infrastructure in European Suburbanization*, Toronto: University of Toronto Press.

McCann, L. 2006. A Regional Perspective on Canadian Suburbanization: Reflections on Richard Harris's 'Creeping Conformity'. *Urban History Review* 35(1): 32–45.

McFarlane, C. 2016. The Geographies of Urban Density: Topology, Politics and the City. *Progress in Human Geography* 40: 629–48.

McFarlane, C., R. Desai and S. Graham. 2014. Informal Urban Sanitation: Everyday Life, Comparison and Poverty. *Annals of the Association of American Geographers* 104: 989–1011.

McFarlane, C. and J. Rutherford. 2008. Political Infrastructures: Governing and Experiencing the Fabric of the City. *International Journal of Urban and Regional Research*, 32, 2: 363–74.

McGee, T. 1991. The Emergence of 'Desakota' Regions in Asia: Expanding a Hypothesis. In N. Ginsberg, ed. *The Extended Metropolis: Settlement Transition in Asia*. Honolulu: University of Hawaii Press, 3–26.

McGee, T. 2013. Suburbanization in a 21st Century. In R. Keil, ed. *Suburban Constellations: Governance, Land and Infrastructure in the 21st Century*. Berlin: Jovis Verlag, 18–25.

McGee, T. 2015. Deconstructing the Decentralized Urban Spaces of the Mega-Urban Regions in the Global South. In P. Hamel and R. Keil, eds. *Suburban Governance: A Global View*. Toronto: University of Toronto Press, 325–36.

McGuirk, J. 2014. *Radical Cities: Across Latin America in Search of a New Architecture*. London: Verso.

Maginn, P. J. and C. Steinmetz, eds. 2015. *(Sub)Urban Sexscapes: Geographies and Regulation of the Sex Industry*. London: Routledge.

Magnusson, W. 2011. *Politics of Urbanism. Seeing like a city*. London and New York: Routledge.

Maher, K. H. 2004. Borders and Social Distinction in the Global Suburb. *American Quarterly* 56(3): 781–806.

Major, C. 2013. Fort McMurray, the Suburb at the End of the Highway. In R. Keil, ed. *Suburban Constellations: Governance, Land and Infrastructure in the 21st Century*. Berlin: Jovis Verlag, 143–9.

Marohn, C. 2015. America's suburban experiment, *Strong Towns Journal*, 15 December 2015; available at http://www.strongtowns.org/curbside-chat-1/2015/12/14/americas-suburban-experiment.

Marohn, C. 2016. Sprawl is not the problem, *Strong Towns Journal*, 18 April 2016; available at http://www.strongtowns.org/journal/2016/4/17/sprawl-is-not-the-problem.

Marshall, C. 2016. A 'radical alternative': how one man changed the perception of Los Angeles, *Guardian*, 24 August, available at https://www.theguardian.com/cities/2016/aug/24/radical-alternative-reyner-banham-man-changed-perception-los-angeles.

Mayer, M. 2012. Metropolitan Research in Transatlantic Perspective: Differences, Similarities and Conceptual Diffusion. In D. Brantz, S. Disko, G. Wagner-Kyora, eds. *Thick Space: Approaches to Metropolitanism*. Bielefeld: Transcript 2012, 105–22.

Mayer, M. 2017. Movements and Politics in the Metropolitan Region. In R. Keil, P. Hamel, J. -A. Boudreau and S. Kipfer, eds. *Governing Cities through Regions: Canadian and European Perspectives*. Waterloo: Wilfrid Laurier University Press, 39–61.

Merrifield, A. 2012. Whither Urban Studies. cities@manchester (http://citiesmcr.wordpress.com/2012/12/10/whither-urban-studies/), 10 December; last viewed, 5 May 2014.

Merrifield, A. 2013. *The Politics of the Encounter. Urban Theory and Protest under Planetary Urbanization*. Athens and London: University of Georgia Press.

Mitchell, D. 2004. *Cloud Atlas*. New York: Random House.

Mitchell, D. 2010. *The Thousand Autumns of Jacob De Zoet*. Toronto: Knopf Canada.

Moench, M. and D. Gyawali. 2008. Desakota: Reinterpreting the Urban-Rural Continuum, Part 2; Re-imagining the Rural-Urban Continuum. Understanding the role ecosystem services play in the livelihoods of the poor in desakota regions undergoing rapid change. Research Gap Assessment by The Desakota Study Team, Institute for Social and Environmental Transition – Nepal (ISET-Nepal).

Monstadt, J. and S. Schramm. 2013. Beyond the Networked City: Suburban Constellations in Water and Sanitation Systems. In R. Keil, ed. *Suburban Constellations: Governance, Land and Infrastructure in the 21st Century*. Berlin: Jovis Verlag, 85–94.

Monte-Mor, R. L. 2014a. Extended Urbanization and Settlement Patterns in Brazil: An Environmental Approach. In N. Brenner, ed. *Implosions / Explosions*. Berlin: Jovis, 109–20.

Monte-Mor, R. L. 2014b. What is the Urban in the Contemporary World? In N. Brenner, ed. *Implosions / Explosions*. Berlin: Jovis, 260–7.

Montero, S. 2016. Study tours and inter-city policy learning: Mobilizing Bogotá's transportation policies in Guadalajara, *Environment and Planning A* 0308518X16669353.

Moore, J. 2014. The Capitalocene, Part I: On the Nature and Origins of Our Ecological Crisis, at: http://www.jasonwmoore.com/Essays.html.

Moos, M. and P. Mendez. 2015. Suburban ways of living and the geography of income: How homeownership, single-family dwellings and automobile use define the metropolitan social space, *Urban Studies*, 52, 10, August, 1864–82.

Moos, M. and R. Walter-Joseph, eds. 2017. *Still Detached and Subdivided*. Berlin: Jovis.

Myerson, Jesse A. 2017. Trumpism: It's Coming From the Suburbs, *The Nation*, 22–9 May. https://www.thenation.com/article/trumpism-its-coming-from-the-suburbs/

Newkirk, V. R. II 2015. The Acts of God. *Seven Scribes*, 19 March; Available at http://sevenscribes.com/the-acts-of-god/.

Nicolaides, B. M. 2002. *My Blue Heaven: Life and Politics in the Working-Class Suburbs of Los Angeles, 1920–1965*. Chicago: The University of Chicago Press.

Nijman, J. 2015. The Theoretical Imperative of Comparative Urbanism: A Commentary on 'Cities beyond Compare?' By Jamie Peck. *Regional Studies* 49(1): 183–6.

Nijman, J. and T. Clery. 2015. The United States: Suburban Imaginaries and Metropolitan Realities. In P. Hamel and R. Keil, eds. *Suburban Governance: A Global View*. Toronto: University of Toronto Press.

Nussbaum, A. 2017. Paris Big Dig Abandons Chic Haussmann for Trains to the Suburbs. Bloomberg, 10 March. https://www.bloomberg.com/news/articles/2017-03-10/paris-big-dig-abandons-chic-haussmann-for-trains-to-the-suburbs.

Ortega, A. 2012. Desakota and Beyond: Neoliberal Production of Suburban Space in Manila's Fringe. *Urban Geography* 33(8): 1118–43.

Ortega, A. 2016. *Neoliberalizing Spaces in the Philippines: Suburbanization, Transnational Migration and Dispossession*. Lexington Books/Rowman & Littlefield.

Osberg, M. 2016. I spent Election Night with the regular-ass suburban white people who elected Trump, Fusion, 10 November, http://fusion.net/story/368805/staten-island-trump-supporters-white-suburbia-despair/.

Park, R. E., E. Burgess and R. McKenzie. 1925. *The City*. University of Chicago Press.

Peck, J. 2015a. Chicago-School Suburbanism. In P. Hamel and R. Keil, eds. *Suburban Governance: A Global View*. Toronto: University of Toronto Press, 130–52.

Peck, J. 2015b. Cities beyond Compare? *Regional Studies* 49(1) (January): 160–82.

Peck, J., E. Siemiatycki and E. Wyly. 2014. Vancouver's suburban involution. *City: Analysis of Urban Trends, Culture, Theory, Policy, Action* 18(4–5). DOI:10.1080/13604813.2014.939464, 386–415.

Phelps, N. A. 2012. The Sub-Creative Economy for the Suburbs in Question. *International Journal of Cultural Studies* 15(3): 259–71.

Phelps, N. A. 2015. *Sequel to Suburbia: Glimpses of America's Post-Suburban Future*. Cambridge: The MIT Press.

Phelps, N. A. ed. 2017. *Old Europe: New Suburbanization? Governance, Land and Infrastructure in European Suburbanization*. Toronto: University of Toronto Press.

Phelps, N. A. and A. T. Vento. 2015. Suburban Governance in Western Europe. In P. Hamel and R. Keil, eds. *Suburban Governance: A Global View*. Toronto: University of Toronto Press, 155–76.

Phelps, N. A. and A. M. Wood. 2011. The New Post-Suburban Politics? *Urban Studies* 48(12): 2591–610.

Phelps, N. A. and F. Wu, eds. 2011. *International Perspectives on Suburbanization: a Post-suburban World?* London: Palgrave Macmillan.

Phillips, M. 2014. Most Germans don't buy their homes, they rent. Here's why. *Quartz Media*, 23 January. http://qz.com/167887/germany-has-one-of-the-worlds-lowest-homeownership-rates/.

Philpott, T. L. 1978. *The Slum and the Ghetto: Immigrants, Blacks and Reformers in Chicago, 1880–1930.* Oxford: OUP.

Pincetl, S., A. E. Jonas and J. Sullivan, 2011. 'Political ecology and habitat conservation for endangered species planning in southern California: region, places and ecological governance'. *Geoforum* 42(4): 427–38.

Pitter, J. 2016. Introduction, In Pitter, J. and J. Lorinc, eds. *Subdivided: City-Building in an Age of Hyper-Diversity.* Toronto: Coach House Books, 5–12.

Pitter, J. and J. Lorinc, eds. 2016. *Subdivided: City-Building in an Age of Hyper-Diversity.* Toronto: Coach House Books.

Poitras, C. 2011. A City on the Move: The Surprising Consequences of Highways. In S. Castonguay and M. Dagenais, eds. *Metropolitan Natures: Environmental Histories of Montreal.* Pittsburgh, PA: University of Pittsburgh Press, 168–86.

Pooley, K. B. 2015. Debunking the 'Cookie-Cutter' Myth for Suburban Places and Suburban Poverty: Analyzing Their Variety and Recent Trends. In K. B. Anacker, ed. *The New American Suburb: Poverty, Race and the Economic Crisis.* Farnham, Surrey: Ashgate, 39–80.

Poppe, W. and D. Young. 2015. The Politics of Place: Place-making Versus Densification in Toronto's Tower Neighbourhoods. *International Journal of Urban and Regional Research.* 39, 3: 613–21.

Pow, C. P. 2012. China Exceptionalism? Unbounding Narratives on Urban China. In T. Edensor and M. Jayne, eds. *Urban Theory Beyond the West: A World of Cities.* London and New York: Routledge, 47–64.

Prigge, W. 1991. Übergänge: Auf der Schwelle einer neuen Stadtentwicklungspolitik. In T. Koenigs, ed. *Vision offener Grünräume: GrünGürtel Frankfurt.* Frankfurt & New York: Campus Verlag, 1173–8.

Pulido, L. 2000. Rethinking Environmental Racism: White Privilege and Urban Development in Southern California. *Annals of the Association of American Geographers* 90(1): 12–20.

Pulido, L. 2015. Geographies of Race and Ethnicity I: White Supremacy vs White Privilege in Environmental Racism Research. *Progress in Human Geography.* 39,6: 809–17.

Quinby, R. 2011. *Time and the Suburbs: The Politics of Built Environments and the Future of Dissent.* Winnipeg: Arbeiter Ring Publishing.

Ranganathan, M. 2014. Paying for Pipes, Claiming Citizenship: Political Agency and Water Reforms at the Urban Periphery. *International Journal of Urban and Regional Research.* 38, 2: 590–608; DOI:10.1111/1468-2427.12028.

Ranganathan, M. and C. Balazs. 2015. Water marginalization at the urban fringe: environmental justice and urban political ecology across the North–South divide. *Urban Geography* 36, 3: 403–23; DOI:10.1080/02723638.2015.1005414.

Ren, X. and R. Keil, eds. 2017. *The Globalizing Cities Reader*. London: Routledge.

Richards, P. M. 2012. *An Integrated Boyhood: Coming of Age in White Cleveland*. Kent: The Kent State University Press.

Rivadulla, M. J. and D. Bocarejo. 2014. Beautifying the Slum: Cable Car Fetishism in Cazucá, Colombia. *International Journal of Urban and Regional Research*, 38, 6 (November): 2025–41.

Robbins, P. 2007. *Lawn People: How Grasses, Weeds and Chemicals Make Us Who We Are*. Philadelphia: Temple University Press.

Robinson, J. 2002. Global and World Cities: A View from off the Map. *International Journal of Urban and Regional Research*, 26, 3: 531–54.

Robinson, J. 2006. *Ordinary Cities: Between Modernity and Development*. London: Routledge.

Robinson, J. 2016. Comparative Urbanism: New Geographies and Cultures of Theorizing the Urban. *International Journal of Urban and Regional Research* 40(1) (January): 187–99.

Robinson, J. and S. Parnell. 2011. Traveling Theory: Embracing Post-Neoliberalism Through Southern Cities. In G. Bridge and S. Watson, eds. *The New Blackwell Companion to the City*. Blackwell Publishing.

Robinson, J. and A. Roy. 2016. Debate on Global Urbanisms and the Nature of Urban Theory. *International Journal of Urban and Regional Research* 40(1) (January): 181–6.

Ronneberger, K. 2015. Henri Lefebvre und die Frage der Zentralität. *Derive*. 60: 23–7.

Ross, A. 2011. *Bird on Fire: Lessons from the World's Least Sustainable City*. Oxford. Oxford University Press.

Ross, B. 2014. *Dead End: Suburban Sprawl and the Rebirth of American Urbanism*. Oxford: Oxford University Press.

Rothstein, R. 2014. *The Making of Ferguson: Public Policies at the Root of its Troubles*. Washington: Economic Policy Institute.

Rousseau, M. 2015. 'Many Rivers to Cross': Suburban Densification and the Social Status Quo in Greater Lyon. *International Journal of Urban and Regional Research* 39, 3: 622–32.

Roy, A. 2009. The 21st-Century Metropolis: New Geographies of Theory. *Regional Studies* 43, 6: 819–30.

Roy, A. 2011. Urbanisms, Worlding Practices and the Theory of Planning. *Planning Theory* 10, 1: 6–15.

Roy, A. 2015. 'Governing the Postcolonial Suburbs'. In P. Hamel and R. Keil, eds. *Suburban Governance: A Global View*. Toronto: University of Toronto Press, 337–48.

Roy, A. 2016. What is Urban about Critical Urban Theory? *Urban Geography* Vol. 37, 6, 810–23.

Roy, A. and A. Ong, eds. 2011. *Worlding Cities: Asian Experiments and the Art of Being Global*. Chichester, West Sussex; Malden, MA: Wiley-Blackwell.

Saunders, D. 2011. *Arrival City: The Final Migration and Our Next World*. Toronto: Vintage Canada.

Saunders, P. 2015. White Urbanization/Black Suburbanization – A Followup. The Corner Sideyard: Yelling from the Window as the World Passes By, 10 November; available at http://cornersideyard.blogspot.ca/2015/11/white-urbanizationblack-suburbanization.html.

Savage, B. 2014. Review. *The Working Man's Reward*, by Elaine Elinek, Oxford: Oxford University Press, 2014. *The Chicago Tribune*, 31 December; available at http://www.chicagotribune.com/lifestyles/books/ct-prj-working-mans-reward-elaine-lewinnek-20141231-story.html.

Savini, F. 2013. Urban Peripheries: The Political Dynamics of Planning Projects. Doctoral dissertation, University of Amsterdam.

Schafran, A. 2013. 'Discourse and dystopia, American style: The rise of 'slumburbia' in a time of crisis'. *City: Analysis of Urban Trends, Culture, Theory, Policy, Action* 17(2). DOI:10.1080/13604813.2013.765125, 130–48.

Schafran, A., O. S. Lopez and J. L. Gin. 2013. Politics and Possibility on the Metropolitan Edge: The Scale of Social Movement Space in Exurbia. *Environment and Planning A* 45: 2833–51.

Schmid, C. 2014. A Typology of Urban Switzerland. In N. Brenner, ed. *Implosions/Explosions*. Berlin: Jovis, 398–427.

Scott, A. J. and E. W. Soja, eds. 1996. *The City: Los Angeles and Urban Theory at the End of the Twentieth Century*. Berkeley: University of California Press.

Semuels, Alana. 2015. Suburbs and the New American Poverty. *The Atlantic*, 7 January; available at http://www.theatlantic.com/business/archive/2015/01/suburbs-and-the-new-american-poverty/384259/.

Seto, K. C., B. Güneralp and L. R. Hutyra. 2012. Global Forecasts of Urban Expansion to 2030 and Direct Impacts on Biodiversity and Carbon Pools. *PNAS early edition* 109(40) (September): 16083–88.

Sevilla-Buitrago, A. 2014. Urbs in rure: Historical Enclosure and the Extended Urbanization of the Countryside. In N. Brenner. ed. *Implosions/Explosions*. Berlin: Jovis, 236–59.

Sewell, J. 1993. *The Shape of the City: Toronto Struggles with Modern Planning*. Toronto: University of Toronto Press.

Sewell, J. 2009. *Shape of the Suburbs: Understanding Toronto's Sprawl*. Toronto: University of Toronto Press.

Sharp, D. and C. Panetta, eds. 2016. *Beyond the Square: Urbanism and the Arab Uprisings*. New York: Urban Research.

Shen, J. 2015. Paving the way for growth: rail transit development and suburbanization in Shanghai, Paper presented at International Forum on Global Suburbanisms, 9–11 May 2015, Fudan University, Shanghai.

Sheppard, E., H. Leitner and A. Maringanti. 2013. Provincializing Global Urbanism: A Manifesto. *Urban Geography*, 34, 7: 893–900, http://dx.doi.org/10.108 0/02723638.2013.807977.

Shields, R. 2012. Feral Suburbs: Cultural Topologies of Social Reproduction, Fort McMurray, Canada. *International Journal of Cultural Studies*, 15, 3: 205–15.

Siemiatycki, M. 2006. Message in a Metro: Building Urban Rail Infrastructure and Image in Delhi, India. *International Journal of Urban and Regional Research* 30(2): 277–92.

Sieverts, T. 2003. *Cities Without Cities: An Interpretation of the Zwischenstadt*. London: Taylor and Francis.

Sieverts, T. 2011. The In-Between City as an Image of Society: From the Impossible Order Towards a Possible Disorder in the Urban Landscape. In D. Young, P. B. Wood and R. Keil, eds. *In-Between Infrastructure: Urban Connectivity in an Age of Vulnerability*. Kelowna: Praxis (e)Press, 19–28.

Sieverts, T. 2015. On the Relations of Culture and Suburbia: How to Give Meaning to the Suburban Landscape? In P. Hamel and R. Keil, eds. *Suburban Governance: A Global View*. Toronto: University of Toronto Press, 239–50.

Simone, A. M. 2004. People as Infrastructure: Intersecting Fragments in Johannesburg. *Public culture* 16(3): 407–29.

Simone, A. M. 2012. No Longer the Subaltern: Refiguring Cities of the Global South. In T. Edensor and M. Jayne, eds. *Urban Theory Beyond the West: A World of Cities*. London and New York: Routledge, 31–46.

Simone, A. M. 2016. It's Just the City After All. *International Journal of Urban and Regional Research*. 40, 1: 210–18.

Skaburskis, A. 2006. New Urbanism and Sprawl: A Toronto Case Study. *Journal of Planning Education and Research* 25(3). DOI:10.1177/0739456X05278985, 233–48.

Smith, N. 2002. New Globalism, New Urbanism: Gentrification as Global Urban Strategy. *Antipode Volume* 34(3) (July): 427–50.

Smith, N. 2006. Foreword. In N.C. Heynen, M.Kaika and E. Swyngedouw, eds. *In the Nature of Cities: Urban Political Ecology and the Politics of Urban Metabolism*. Abingdon: Routledge, xi–xv.

Smith, P. D. 2012. *City: A Guidebook for the Urban Age*. London: Bloomsbury.

Smith, A. C. and N. Bedi. 2016. Trump won Florida on strength of suburban white vote, *Tampa Bay Times*, 9 November; available at http://www.miamiherald.com/news/politics-government/election/article113786328.html.

Soja, E. W. 1996. *Thirdspace: Journeys to Los Angeles and Other Real-and-Imagined Places*. Cambridge, Massachusetts: Blackwell.

Soja, E. W. 2000. *Postmetropolis: Critical Studies of Cities and Regions*. Oxford: Blackwell.

Soja, E. W. 2010. *Seeking Spatial Justice*. Minneapolis, University of Minnesota Press.

Stanilov, K. 2007. Housing Trends in Central and Eastern European Cities During and after the Period of Transition. In K. Stanilov, ed. *The Post-Socialist City: Urban Form and Space Transformations in Central and Eastern Europe after Socialism*. Dordrecht: Springer Verlag, 173–90.

Stanilov, K. and L. Sykora, eds. 2014a. *Confronting Suburbanization: Urban Decentralization in Postsocialist Central and Eastern Europe*. Oxford: Wiley-Blackwell.

Stanilov, K. and L. Sykora. 2014b. Managing Suburbanization in Postsocialist Europe. In K. Stanilov and L. Sykora, eds., 2014. *Confronting Suburbanization: Urban Decentralization in Postsocialist Central and Eastern Europe*. Oxford: Wiley-Blackwell, 296–320.

Steffen, W., P. J. Crutzen and J. R. McNeill. 2007. The Anthropocene: Are Humans Now Overwhelming the Great Forces of Nature? *Ambio* 36(8): 614–21.

Stone, C. N. and H. T. Sanders. 1987. *The Politics of Urban Development*. Lawrence: University Press of Kansas.

Story, L. and S. Saul. 2015. Stream of Foreign Wealth Flows to Elite New York Real Estate. *New York Times Dossier*. Available at http://www.nytimes.com/2015/02/08/nyregion/stream-of-foreign-wealth-flows-to-time-warner-condos.html?_r=0.

Stromberg, J. 2016. Highways gutted American cities. So why did they build them? *Vox. com* 11 May; Available at http://www.vox.com/2015/5/14/8605917/highways-interstate-cities-history.

Swilling, M. 2016. The curse of urban sprawl: how cities grow and why this has to change, *Guardian*, 12 July, available at https://www.theguardian.com/cities/2016/jul/12/urban-sprawl-how-cities-grow-change-sustainability-urban-age.

Swyngedouw, E. 1996. The City as Hybrid: On Nature, Society and Cyborg Urbanization. *Capitalism Nature Socialism* 7(2): 65–80.

Swyngedouw, E. 2004. *Social Power and the Urbanization of Water: Flows of Power*. Oxford: Oxford University Press.

Swyngedouw, E. 2011. Depoliticized Environments: The End of Nature, Climate Change and the Post-Political Condition. *Royal Institute of Philosophy Supplement* 69: 253–74.

Swyngedouw, E. and M. Kaika. 2014. Urban Political Ecology. Great Promises, Deadlock...and New Beginnings? *Documents d'Analisi Geografica* 60(3): 459–81.

Sykora, L and K. Stanilov. 2014. The Challenge of Postsocialist Suburbanization. In K. Stanilov and L. Sykora, eds. 2014a. *Confronting Suburbanization: Urban Decentralization in Postsocialist Central and Eastern Europe*. Oxford: Wiley Blackwell, 1–32.

Szirmai, V. ed. 2011. *Urban Sprawl in Europe: Similarities or Differences?* Budapest: AULA Kiadó.

Tartt, D. 2013. *The Goldfinch*. New York: Little Brown and Company.

Teaford, J. 1997. *Post-Suburbia: Government and Politics in the Edge Cities*. Baltimore: Johns Hopkins University Press.

Teaford, J. 2011. Suburbia and Post-Suburbia: A Brief History. In N. A. Phelps and F. Wu, eds. *International Perspectives on Suburbanization: A Post-Suburban World?* London: Palgrave Macmillan, 15–34.

Tonkiss, F. 2013. *Cities by Design: The Social Life of Urban Form*. Cambridge: Polity.

Touati-Morel, A. 2015. Hard and Soft Densification Policies in the Paris City-Region. *International Journal of Urban and Regional Research* 39, 3: 603–12.

Tzaninis, Y. 2016. *Building Utopias on Sand: The Production of Space in Almere and the Future of Suburbia* (Doctoral dissertation). University of Amsterdam.

Ucoglu, Murat. 2016. Istanbul's Suburban Dream Is Fueled by Debt, *CityLab*, 25 February; http://www.citylab.com/housing/2016/02/istanbuls-suburban-dream-is-fueled-by-debt/470550/.

United Nations, Department of Economic and Social Affairs, Population Division. 2014. *World Urbanization Prospects: The 2014 Revision*, Highlights (ST/ESA/SER. A/352).

United Way of Toronto. 2011. *Vertical Poverty. Poverty by Postal Code 2*. Toronto: United Way.

Unless We Build Them Right, Time Magazine, 18 September, available at http://science.time.com/2012/09/18/urban-planet-how-growing-cities-will-wreck-the-environment-unless-we-build-them-right/.

Urry, J. 2014. Epilogue. In P. Adey, D. Bissell, K. Hannam, P. Merriman and M. Sheller, eds. *The Routledge Handbook of Mobilities*. London: Routledge, 585–92.

Vaughan, L., S. Griffiths, M. Haklay and C. E. Jones. 2009. Do the Suburbs Exist? Discovering the Complexity and Specificity in Suburban Built Form. *Transactions of the Institute of British Geographers* 34 (4): 475–88.

Veracini, L. 2011. Suburbia, Settler Colonialism and the Word Turned Inside Out. *Housing, Theory and Society* 29(4): 339–57.

Vicino, T. J. 2008. *Transforming Race and Class in Suburbia: Decline in Metropolitan Baltimore.* New York: Palgrave Macmillan.

Wachsmuth, D. 2012. Three Ecologies: Urban Metabolism and the Society–Nature Opposition. *The Sociological Quarterly* 53(4): 506–23.

Wacquant, L. 2008. *Urban Outcasts: A Comparative Sociology of Advanced Marginality.* Malden: Polity.

Wagner, A. 1935. *Los Angeles: Zweimillionenstadt in Südkalifornien.* Leipzig.

Waldie, D. J. 1996. *Holy Land: A Suburban Memoir.* Los Angeles: W. W. Norton.

Walker, R. 1977. The Suburban Solution: Capitalism and the Construction of Suburban Space in the United States. Ph.D. Dissertation, Johns Hopkins University.

Walker, R. 1981. A Theory of Suburbanization: Capitalism and the Construction of Urban Space in the United States. In M. Dear and A. Scott, ed. *Urbanization and Urban Planning in Capitalist Societies.* New York: Methuen, 383–430.

Walker, R. and R. Lewis. 2004. Beyond the Crabgrass Frontier: Industry and the Spread of North American Cities, 1850–1950. In R. Lewis, ed. *Manufacturing Suburbs: Building Work and Home on the Metropolitan Fringe.* Philadelphia: Temple University Press, 16–31.

Walker, R. and R. D. Lewis. 2001. 'Beyond the Crabgrass Frontier: Industry and the Spread of North American Cities, 1850–1950.' *Journal of Historical Geography* 27(1): 3–19.

Walks, A. 2013. Suburbanism as a Way of Life, Slight Return. *Urban Studies* 50(8): 1471–88.

Walsh, B. 2012. Urban Planet: How Growing Cities Will Wreck the Environment Unless We Build Them Right, *Time Magazine,* 18 September, available at http://science.time.com/2012/09/18/urban-planet-how-growing-cities-will-wreck-the-environment-unless-we-build-them-right/.

Warner, S. B. 1978. *Streetcar Suburbs: The Process of Growth in Boston (1870–1900).* Cambridge: Harvard University Press.

Warner, S. B. 1972. *The Urban Wilderness: A History of the American City.* Berkeley and Los Angeles: University of California Press.

Weber, M. 1921. Die nichtlegitime Herrschaft (Typologie der Städte), in *Wirtschaft und Gesellschaft.* Tübingen: J. C. B. Mohr (Paul Siebeck), 1976.

Webster, R., ed., 2000. *Expanding Suburbia: Reviewing Suburban Narratives.* New York and Oxford: Berghahn Books.

Whitehead, M. 2014. *Environmental Transformations.* London: Routledge.

Wiese, A. 2004. *Places of Their Own: African American Suburbanization in the Twentieth Century.* Chicago: The University of Chicago Press.

Wu, F. 2013. Chinese Suburban Constellations: The Growth Machine, Urbanization and Middle Class Dreams. In R. Keil, ed. *Suburban Constellations: Governance, Land and Infrastructure in the 21st Century*. Berlin: Jovis Verlag, 190–4.

Wu, F. and J. Shen. 2015. Suburban Development and Governance in China. In P. Hamel and R. Keil, eds. *Suburban Governance: A Global View*. Toronto: University of Toronto Press, 303–24.

Wu, F. and N. A. Phelps. 2008. From suburbia to post-suburbia in China: aspects of the transformation of the Beijing and Shanghai global city regions. *Built Environment* 34(4): 464–82.

Yiftachel, O. 2009. Theoretical Notes On 'Gray Cities': the Coming of Urban Apartheid? *Planning Theory*, vol. 8, 1: 88–100.

Young, D. and R. Keil. 2014. Locating the Urban In-between: Tracking the Urban Politics of Infrastructure in Toronto. *International Journal of Urban and Regional Research* 38(5): 1589–608.

Young, D., P. B. Wood and R. Keil, eds. 2011. *In-Between Infrastructure: Urban Connectivity in an Age of Vulnerability*. Kelowna: Praxis (e)Press.

Zumelzu, A. 2011. La Sustentabilidad como problema mundial en la era del carbon Eindhoven como ejemplo de ciudad antropocena. *AUS (Valdivia)*, 2011, no. 10, 4–7. ISSN 0718-7262.

Index

Page numbers **in bold** refer to photographs.

access restrictions. *See* gated or concierged communities
Adorno, Theodor W., 74
Africa: country-to-city migration, 37, 123; density, 5, 54, 158; diseases and suburbanization, 175; diverse processes and forms, 156; habitable and uninhabitable spaces, 170–1; infrastructure, 139; scholarship on, 71; squatter settlements, 37, 158; trends, 1, 5, 18. *See also* Global South
African Americans: gentrification's impact on, 85, 103, 104; home ownership, 32; marginal settlement spaces, 105–7; pejorative discourse, 104, 198; policing, 105, 182; popular culture, 103–4; poverty, 32, 102–3; racism, 32, 96, 104–7, 203n7; *The Sellout* (Beatty), 203n6. *See also* Ferguson, Missouri; race and ethnicity; United States
agricultural land: global suburbanization, 128, 171–2; greenbelts, 165–6; historical background, 27; infrastructure, 132, 168; para-agricultural land, 166; pastoral capitalism, 171–2. *See also* natural environment
Aguilar, Adrián G., 124
Ahmed-Ullah, Noreen, 100
airports, 38, 132, 142
Almere, Netherlands, 87–8, 93
Amadora, Portugal, 127, **127**
Amin, Ash, 125
Amsterdam, 37
Anatomy of Los Angeles (film), 91–2
Anderson, Elijah, 96
Angel, Shlomo, 4
Anglo-Saxon societal model, 110–11. *See also* settler societies; single-family home suburb
Anthropocene, 153, 173–5, 177–9. *See also* environmental sustainability
Anting, China, 145, **180**
Arboleda, Martin, 173, 178
Arcade Fire, 101
arrival cities, 4, 37, 99, 177, 193. *See also* immigrants
Arrival Cities (Saunders), 37
Asia: country-to-city migration, 37, 123; *desakota* (city/rural links), 123–4; inter-referencing, 145–6; trends, 18. *See also* China; Global South; India

INDEX   235

Asian Americans, 98
Australia: as settler society, 110–11; single-family home suburb, 13, 29–30
automobility: about, 30–1, 133; commuter patterns, 147; 'drive till you qualify' home ownership, 32; environmental sustainability, 74–5, 140–1, 144, 203n2 (ch. 7); global suburbanization, 139; historical background, 30–1, 33, 147; Los Angeles as model, 91; poverty in car-centric suburbs, 96; retrofitting, 74–5; symbol of freedom, 30–1; technological change, 144; transit oriented development (TOD), 144–5; trends, 144, 148. *See also* infrastructure; mobility

Bain, Alison, 73–4
Balducci, Alessandro, 148–9
Banham, Reyner, 91
Barraclough, Laura, 73, 165
Baum-Snow, Nathaniel, 33
Beatty, Paul, 203n6
Beauregard, Robert, 114, 186–7
Beijing, 156
Belgrade, Serbia, 127, **128**
Belo Horizonte, Brazil, **2**, 166
Benjamin, Walter, 165–6
Birmingham, UK, 118–19, 165
Blacks. *See* African Americans
Bogota, Colombia, 145
boundaries: about, 163–7, 197; call for new theory, 197; extended urbanization, 177–9; global suburbanization, 164–5; governance and politics, 164, 185, 197; habitable and uninhabitable, 170–1; historical background, 7, 164, 172; mobility's impact on, 133, 137; natural environment, 163–4; post-suburbanization, 164–5, 168–9; real and imagined landscapes, 172; regions, 185; suburbanization as boundary setting, 163–4; as thresholds, 165–6; UPE (urban political ecology), 163, 167–8; urban vs. rural, 164, 172. *See also* greenbelts; mobility
Brampton, Ontario, 100
Braudel, Fernand, 70
Brazil: country-to-city migration, 123; diseases and suburbanization, 175; diversity of processes and forms, 115, 116; gated communities, 146; governance and politics, 141–2, 182, 188; high-rise developments, 37, 114, 115, 116; informal settlements, 115, 146, 188; infrastructure inequities, 141–2, 146; mobility, 115, 145–6; modernism, 114; poverty, 141–2, 146
Britain. *See* United Kingdom
Brown, Michael, 84–5, 101–2, 108. *See also* Ferguson, Missouri
bungalow, 111. *See also* single-family home suburb
Bunnell, Tim, 85

California: boundaries, **82**; centre/periphery dynamics, 185; governance and politics, 89–90,

California: boundaries (cont.) 182; inequities, 89–90; mobility, 147–8; race and class, 89–90; technoburbs, 57, 78, 147–8. *See also* Lakewood, California; Los Angeles area

Canada: about, 116–17; cultural life in suburbs, 73; density, 157; high-rise developments, 94–5, 117, 157; historical background, 29–30, 116–17; majority minority suburbia, 97–100; political shortchanging of suburbs, 184; scholarship on, 67–8, 116–17; self-built homes, 116–17; as settler society, 110–11; single-family home suburb, 29–30, 116–17. *See also* single-family home suburb

capitalism: about, 15–16, 19–20, 24, 173–4, 197; call for new approaches, 199; capital accumulation, 140; capital switching, 24, 29, 197; capitalocene, 174, 179, 199; Dejima as early global suburb, 86–8, 93; environmental sustainability, 173–4; extended urbanization, 175–6; global suburbanization, 127–8; governance and politics, 184, 197; historical background, 29–30, 63; home ownership, 30–2; inequities, 173; infrastructure development, 140; as integral to suburbanization, 64–5; land development, 19–20, 24, 28, 173–5, 184, 196–7; ostentatious displays of wealth, 38, 56–7, 140; pastoral capitalism, 171–2; privatism, 76; urban theory, 24. *See also* consumerism; creative sectors in urban centre; home ownership; neoliberalization

capitalocene, 174, 179

cars. *See* automobility

Carver, Humphrey, 111

Castells, Manuel, 5–6, 48, 64

centre/periphery dynamics: about, 14; call for new theory on, 50–1, 62–3, 80; centrifugal city, 10–11, 114–16, 121–2; as continuum, 19; Global South, 80; governance and politics, 185; inbetween cities, 75; interdependence, 8–9; in Los Angeles, 90–2; new centralities, 50, 55, 189; sub/urban studies, 64; terminology, 15; trends, 177. *See also* explosion/implosion dialectics; gentrification; suburbanization

Chakrabarti, Vishaan, 159

change and continuity: assumption of change, 197; dialectics of, 93; environmental sustainability, 178, 193; suburbs as capable/incapable of change, 161, 193; technological change and infrastructure, 134, 137, 144, 145

Charmes, Eric, 161

Cheng, Wendy, 98

Chicago School, 14, 47, 64, 91

Chile, mining towns, 177, 178, 198

China: about, 128–9; diversity of processes and forms, 128–9, 145, 156, **180**, 196; high-rise developments, 5, 37, 51, 54, 123, 175, 176; infrastructure, 129, 139–40; mining sand for concrete, 175–6; new towns,

128–9, 145, **180**; satellite cities, 144–5; scholarship on, 71; transit oriented development (TOD), 144–5; trends, 5, 176, 196
Chinese Americans, 98
choice, individual, 24–5, 68, 190–1
City Institute at York University, 44
civil society, 142, 184. *See also* governance and politics
climate change: about, 173–6; call for new approaches, 199–200; capitalism and production of urban space, 173–5; change to counter, 178, 193; failed vs. successful suburbanization, 53–4; historical background, 174–5; metabolisms, 152–3; retrofitting, 74–6; sprawl, 140–1; trends, 4–5; urban political ecology (UPE), 151–3. *See also* environmental sustainability
Clinton, Hillary, 181
Cloud Atlas (Mitchell), 16–19
communications, 137, 142–3, 148–9. *See also* infrastructure
compactness. *See* density
Compton, California, 103–4
'Compton' (song), 103–4
concierged communities. *See* gated or concierged communities
Congo, 170–1
consumerism: about, 29–30; driver of urbanization, 20; Fordist economy, 24, 155; malls, 29, 193. *See also* capitalism; Fordist and post-Fordist economy
continuity. *See* change and continuity

Cowen, Deborah, 85, 104–5
Crawford, Margaret, 77–8, 97–8
creative sectors in urban centre: about, 25–6, 197; gentrification, 58; millennials, 26, 148; monocultures, 58; re-urbanization, 10; suburbs as 'sub-creative' place, 26. *See also* gentrification
cultural life in suburbs: about, 73–4; conflict management, 88, 98; conformity, 26, 30–1, 74; continuity and change, 75–6, 93–4, 161, 193; diversity vs. homogeneity, 90–1, 97–8; dystopia vs. utopia, 28, 62, 101, 104; greenbelts, 168–9; historical background, 28; identity of place, 103–4; metabolisms, 168–9; myth of safety, 105; normative ideals, 62, 77; patriotic narratives, 110; populism and protests, 181–3; privacy, 30–1; retrofitting, 75; scholarship on, 73–4; white suburbs as political mindset, 181–2. *See also* everyday life; natural environment; popular culture and arts
Cultures of the Suburbs project, 73

Das neue Frankfurt, 113–14
Davis, Mike, 55, 66, 101, 182
De Jong, Judith, 74
De Meyer, Dirk, 194
de-centralization. *See* centre/periphery dynamics
decline of suburbs: about, 65–6, 73, 96–7; diversity in, 95–6; 'end of suburbia,' 73, 96; mobility, 96; pejorative discourse, 96–7;

decline of suburbs: about (cont.)
post-subprime suburbia, 66–7, 96; poverty, 94–7; scholarship on, 73, 97. *See also* retrofitting; re-urbanization
definitions. *See* terminology
de-industrialization: historical background, 18, 34; portrayal in *Cloud Atlas*, 16–18. *See also* industrial and commercial suburbanization
Dejima, Japan, 86–8, 93
Delhi, 127, 146, 165, 171–2
demographics: diversification in Almere, 87–8; Global South, 49; poverty in US suburbs, 94; trends, 3–5, 174
Demos, T. J., 173–4
density: about, 4, 153–4, 161–3, 191–2; assumptions in debates on, 160–3, 172; call for new approaches, 199; critical questions on, 157; debates on compactness vs. sprawl, 153–4, 156–63, 172, 175–6, 190–2; diversity of processes and forms, 154–6, 158, 161–3, 172, 190–1; environmental sustainability, 4, 76, 151–3, 156, 159, 160, 172, 191; explosion/implosion dialectics, 6–7; global suburbanization, 5, 155–6, 158, 160–1, 172, 175–6; governance and politics, 161, 190–2; hyperdensity, 154, 191; infrastructure's impact, 135, 156; metabolisms, 152–3; normative ideals, 75, 153–4, 156–63, 172, 175–6; poverty, 159; public park vs. private garden, 163; retrofitting, 75–6; single-family home model, 153–6; social class, 159–63, 191; suburb as part of larger system, 152–3, 156, 160, 163; terminology, 157–8; trends, 4, 5, 15–16, 117–18; urban planning, 158, 160, 172, 175–6. *See also* sprawl
desakota (city/rural links), 123–4
Dikeç, Mustafa, 141
diseases and suburbanization, 175, 177
displacement: about, 33–4; centrifugal city, 114–15; distanciation strategy, 34, 47–8, 65; fear of the Other, 65; as global strategy, 33; Haussmannization, 33, 183; historical background, 28. *See also* exclusion; gentrification
disposition, 112, 139–40, 143, 186
diversity: about, 38–9, 49–50, 95–7, 188, 198–9; and contested global processes, 85; gentrification, 183; global suburbs, 12–13, 49–50; governance and politics, 39, 100, 183; historical background, 66; majority minority suburbia, 97–101, 104; primary urbanization, 49; of processes and forms, 42–3, 49–52, 54, 66, 75; segregation in USA, 95–6; socio-economic, 96–7; of suburban processes and forms, 11, 13; worlding, 20, 49, 80. *See also* explosion/implosion dialectics;

immigrants; race and ethnicity; social class
Dreier, Peter, 102–4, 106–7
Drummond, Lisa, 70–1, 196
Dublin, Ireland, 118
Dunham-Jones, Ellen, 74, 143

Easterling, Keller, 110, 112, 139–40, 143, 155, 186
Eastern Europe. *See* Europe, Eastern
Ebola virus, 175
economy: post-Fordist economy, 37–8, 57, 99; service economy, 48, 58, 99. *See also* capitalism; Fordist and post-Fordist economy; neoliberalization
Ehrenhalt, Alan, 18, 94
Ekers, Michael, 15
England. *See* United Kingdom
environmental sustainability: about, 152–3; Anthropocene, 173–5, 177–9; call for new approaches, 8, 199–200; change potential, 178, 193; density debates, 156–63, 172, 175–6, 191; extended urbanization, 8, 177–9; failed vs. successful suburbanization, 53–4; governance and politics, 141, 161; metabolisms, 152–3, 176; mobility, 74–5, 134, 203n2 (ch. 7); normative ideals, 77, 191; pollution in industrial suburbs, 25, 28, 38, 105–6; population projections, 174; retrofitting, 74–6, 143; scholarship on, 152; suburb as part of larger system, 136–7, 152, 156; trends, 4–5, 15–16; urban political ecology (UPE), 151–3,
163. *See also* climate change; natural environment
Ernstson, Henrik, 176
Estonia, **150**
ethnicity. *See* race and ethnicity
ethnoburbs: arrival cities, 4, 37, 99, 177, 193; as 'cities-in-waiting,' 57; community structures, 97, 193; immigration, 38; Los Angeles area, 98–9; majority minority suburbia, 99–101; scholarship on, 97. *See also* immigrants; race and ethnicity
Euro-American model of suburbs. *See* single-family home suburb
Europe: about, 35–6, 117–18; collective agency for regions, 184; crisis in housing estates, 36, 117, 121; diversity of processes and forms, 117–18, 127, **127**, 187; environmental sustainability, 145; gentrification, 203n5; governance and politics, 184; historical background, 18, 35–6; home ownership, 31; housing estates, 35–6, 117, 120–1, **150**, 187; mobility, 121, 145; modernism, 113–14; re-urbanization, 36, 117–18, 121; scholarship on, 71; state-led or co-op development, 36, 117, 121, 187; trends, 5, 117–18. *See also* France; Germany; Netherlands
Europe, Eastern: about, 35–6, 117–18, 121–2; centrifugal cities, 121–2; crisis in housing estates, 36, 117, 121; diversity of processes and forms, 117–18, 121–2, 127, **128**, 155; gated

Europe, Eastern: about (cont.)
communities, 121–2; governance and politics, 188, 192; historical background, 35–7; home ownership, 31; housing estates, 35, 117, 120–1, **150**, 187–8; mobility, 121; post-suburbanization, 155; privatization, 36–7, 121; re-urbanization, 117–18, 121; scholarship on, 71; separation of land uses, 120–1; sprawl, 37, 121; state-led or co-op development, 117, 121, 187–8; urban regeneration, 37. *See also* Germany, East

everyday life: about, 6, 177; 'cityness' as space of encounters, 193; continuity and change, 75–6, 93–4, 161, 193; environmental sustainability, 177; ethnographic research, 70–1; extended urbanization, 8, 177–9; gentrification, 183; governance and politics, 183, 188–9; malls, 29, 193; mobility, 135; retrofitting, 193; suburban as success symbol, 8; suburbanism, as term, 11. *See also* cultural life in suburbs; mobility; normative ideals; popular culture and arts

exclusion: about, 33–4; centrifugal city, 114–15; distanciation strategy, 34, 47–8, 65; ethnoburbs, 98–9; fear of the Other, 65; forms of segregation, 55; as global strategy, 33; grey spaces, 169–70; Haussmannization, 8, 33, 183; historical background, 28; infrastructure inequities, 141; Los Angeles area, 73, 98–9; racialized public policies, 106–7; spaces of privilege, 83; systemic racism, 106–7; unwanted functions, 28. *See also* displacement

explosion/implosion dialectics: about, 10, 52–3, 69–70; call for new theory on, 50–1; centralist tradition, 58–9, 70; diversity of, 43–4, 55, 70; fragments, 43, 70, 116, 125; global suburbanization theory, 52–3, 69–70, 138; growth dynamics, 6–7; infrastructure, 138, 148–9; Lefebvre on, 3, 6, 10, 27, 43, 52, 69–70, 138, 196; mobility, 148–9; unknowability of, 7

extended urbanization, 8, 49–50, 175, 177–9. *See also* suburbanization

exurbs: airports, 38; impact of 2008 financial crisis, 96; ostentatious displays of wealth, 38; poverty, 96

factory production. *See* Fordist and post-Fordist economy

families: choice of residence, 158–9, 195; diversity of, 158–9, 195; explosion/implosion dialectics, 148–9

Ferguson, Missouri: about, 84–5, 101–5; gentrification of St. Louis, 85, 104; governance and politics, 85, 105, 107;

marginalizing language, 198; racism, 85, 107
Filion, Pierre, 134–5
Filler, Martin, 76–7
film and TV portrayals, 74, 90–2, 202n2 (ch. 5)
financial crisis (2008): impact on home ownership, 31, 106; impact on suburbia, 66–7; post-subprime suburbia, 96
Fishman, Robert, 65–7, 70, 78, 79, 113, 114
Fleischer, Friederike, 156, 195–6
Flevopolder (artificial island), 87–8, 93
Florida, Richard, 26
Fogelson, Robert, 65
Fordist and post-Fordist economy: about, 29–30; capital switching, 24, 197; consumerism, 24; global suburbanization, 197; historical background, 18, 29–30; home ownership, 31; neoliberalization, 56; post-Fordist economy, 37–8, 57, 99; production of single-family homes, 29–30, 33, 112; suburbanization and climate change, 174–5
Fort McMurray, Canada, 178
France: about, 119; 'decision-making centres,' 183; density, 192; displacement from core, 35, 183; governance and politics, 182, 192; Haussmannization, 33, 183; housing estates, 35, 117, 141; infrastructure, 141; mobility, 147; right to the city, 141, 183; single-family home subdivisions, 35; social class, 35; social exclusion, 141. See also Europe; Paris
Frankfurt: greenbelts, 165; growth dynamics, 6–7; housing estates, 35, 113–14, 119, 120. See also Europe; Germany
Friedman, Andrew, 187
Friedmann, John, 7

Gans, Herbert, 202n1 (ch. 5)
Garner, Eric, 182
gated or concierged communities: density, 159; as distanciation strategy, 34; environmental sustainability, 53, 159; forms of segregation, 55; global suburbanization, 37, 121–3; primary urbanization, 49; scholarship on, 68; wealth and class, 55, 159
gender and modernization, 112–13
gentrification: about, 10, 33, 183; African Americans, 85, 103, 104; displacement and exclusion, 148, 159, 177, 183; as distanciation strategy, 47–8; global suburbanization, 33, 124; Haussmannization, 8, 33, 183; mobility, 148, 159; monocultures, 58; protests, 183; service economy, 58; St. Louis's impact on Ferguson, 85, 104; urban elites, 29
Germany: about, 35–6, 119–20; diversity of processes and forms, 119–20; greenbelts, 167; high-rise developments, 120; historical background, 35–6, 119–20; home ownership, 31,

Germany: about (cont.) 36; housing estates (*Großwohnsiedlung*), 35, 117, 120; modernism, 119–20; single-family home suburbs, 35–6, 119–20; social class, 35. *See also* Europe; Frankfurt

Germany, East: about, 120–2; housing estates, 35, 117, 120–1; *Plattenbau*, 117, 121; re-urbanization, 37, 121; separation of land uses, 120–1; sprawl, 37, 121. *See also* Europe, Eastern

Get Out (film), 73

Global North: privileging in urban studies, 16, 58. *See also* Canada; Europe; Europe, Eastern; metabolisms; United Kingdom; United States

Global South: call for new theory on, 80; country-to-city migration, 37, 68, 123; demographics, 49; density, 158; *desakota* (city/rural links), 123–4; diasporic cultures, 123; diseases and suburbanization, 175, 177; diversity of processes and forms, 122–3, 125, 127–8, 196; dystopia vs. utopia, 125; extended urbanization, 49; historical background, 37; influence on theory, 67–8, 123–4; informal settlements, 188; infrastructure inequities, 141–2, 146; innovations, 145; inter-referencing, 42, 126, 145–6, 158; mobility, 139, 145–6; scholarship on, 67–8, 71; terminology, 125–6. *See also* Africa; Asia; China; India; Latin America; metabolisms; South East Asia

Global Suburbanisms (York University), 15, 44–5, 69

global suburbanization: about, 3–5, 9–16, 86–8, 122–9, 199; bias of Euro-American model, 16, 62–3, 126; boundaries, 164–5; centralist tradition, 198; demographic mixing, 86, 87–8; density debates, 156; diversity of processes and forms, 12–13, 42–3, 48–50, 54–5, 115–16, 122–3, 127–8; dystopia vs. utopia, 52; elite escapism, 114–15, 121–2; environmental sustainability, 164, 172, 199; ethnographic research, 70–1; extended urbanization, 8, 15–16, 49–50, 177–9; external forces on, 115–16, 122–3; failed vs. successful suburbanization, 53–4; as focus of this book, 10–16, 19; fragments, 116, 125; gentrification, 33; global cities, 38, 48, 85–6; governance and politics, 54–5; historical background, 86–8, 93; infrastructure, 136–8, 144–7; intense development, 127–8; inter-referencing, 42, 115–16, 126, 145–6, 158; mixed land-use, 127–8; modernism, 113–14; modernization, 112–13; multiple centralities and peripheralities, 16, 50, 53, 55, 177; poverty, 115; primary urbanization, 49; scholarship on, 71; suburb as part of larger system, 126, 136–7; terminology,

45–6; trends, 3–8, 15–16, 18, 20; unknowability of, 7, 16–19, 58; urban planning, 172. *See also* explosion/implosion dialectics; Global South; metabolisms; terminology; theory and global suburbanization
The Goldfinch (Tartt), 109–10
Gottmann, Jean, 194
governance and politics: about, 39, 184–5, 198–9; authoritarian privatism, 184–5; call for new approaches, 184, 198–200; capitalist accumulation, 184–5; centralist tradition, 23, 39; civil society, 142, 184; collective agency, 184; decision making, 135, 141; density debates, 161–3, 190–2; dystopia vs. utopia, 101; environmental issues, 161; formal and informal approaches, 188–9; gentrification, 85; global suburbanization, 54–5; historical background, 185; inequities and justice, 142, 191; infrastructure, 134–5, 141–2; intersectionality, 100; intervention vs. privacy, 194; neoliberalism, 183–4, 190; 'parasitic suburbs,' 186–7; policing, 101, 105, 182; populism and protests, 181–3; post-suburbanization, 189–90; regime theory, 184; regions, 39, 184–5, 187; retrofitting, 192–4; scholarship on, 184; segregation, 89–90; shortchanging of suburbs, 184; state policy, 184–5; suburban and rural votes in US, 181–2

Graham, Stephen, 142–3
Great Britain. *See* United Kingdom
greenbelts: about, 164–70; density, 76; global suburbanization, 164–5, 167; grey spaces, 169–70; historical background, 166–7; leapfrogging, 165; metabolisms, 167, 168–9; neoliberal economic value, 168; Ontario, 7, **162**; post-suburbanization, 168–9; real and imagined landscapes, 166, 168; as thresholds, 165–6; UPE (urban political ecology), 167–8. *See also* environmental sustainability; natural environment
Gregory, Derek, 20
grey spaces, 169–70
Gurgaon, India, 127, 165, 171–2
Gururani, Shubhra, 127, 165, 171–2

Hamel, Pierre, 15
Harris, Richard, 23, 30–1, 46, 67–8, 72, 78, 116–17
Harvey, David, 11, 55
Häußermann, Hartmut, 185
Haussmannization, 8, 33, 183
Hayden, Dolores, 65–6
Herzog, Lawrence, 114, 116
high-rise developments: elite escapism, 114–15; global suburbanization, 176; mining sand for concrete, 175–6; poverty, 94–5, 115; primary urbanization, 49; stranded areas, 157; symbol of decline, 94–5, 120, 191
Hirt, Sonia, 155, 187

history of sub/urbanization, pre-20th c.: about, 27–9, 174; Anglo Saxon societal model, 110–11; boundaries, 164, 172; capitalist societies, 86; colonialism, 111; Dejima as early global suburb, 86–8, 93; displacement and exclusion, 28; extended urbanization, 8; global suburbs, 86–8, 93; military defense, 27, 164; 'new land,' 27–8

history of sub/urbanization, 20th c.: about, 29–30; centralist tradition, 57–9, 63; displacement, 33–4; greenbelts, 166–7; home ownership, 30–2; housing estates in Europe, 35–6; industrialization, 32–3; mobility, 33; modernism, 113–14. *See also* explosion/implosion dialectics; Fordist and post-Fordist economy; Lefebvre, Henri; single-family home suburb

Hoch, Charles, 89
Hofstra University, 72
Holy Land (Waldie), 77–8, 89–90, 93
home ownership: about, 30–2; Euro-American model of suburbs, 30–2; by immigrants, 30, 99, 106; normative ideal, 77; privacy, 30–1, 65, 194; racialized policies, 105–7; symbol of freedom, 30
Hong Kong, 37
Horkheimer, Max, 74
housing: architecturally polyglot peripheries, 128–9; designs as multipliers, 155; diversity of, 155–6; global suburbanization, 128–9; governance and politics, 187–9; mobile homes, 188–9; modernism, 113–14; multi-family housing, 195; retrofitting, 143; statistics on US housing, 159; suburban forms, 203n1 (ch. 6). *See also* consumerism; high-rise developments; self-built settlements; single-family home suburb; squatter settlements

Huasco, Chile, 177, 178, 198
Hudalah, Delik, 124
Huq, Rupa, 73

ideals. *See* normative ideals
identity of place, 103–4
immigrants: arrival cities, 4, 37, 88, 99, 162, 177, 193; conflict with neighbours, 88; country-to-city, 4, 37, 68, 123; global suburbs, 48, 86–8, 93; home ownership, 30, 32, 106; inter-referencing, 115–16. *See also* ethnoburbs
Imperial Valley, California, 182
inbetween cities: about, 18, 75, 78–9; density debates, 157; *desakota* (city/rural links), 123–4; displacement and exclusion, 28; dystopia vs. utopia, 52; greenbelts, 167; historical background, 18, 28; processes, 75, 79–80; terminology, 45
India: agricultural past, 171; density, 5; diversity of processes and forms, 127–8, 156;

INDEX 245

greenbelts, 165; high-rise developments, 37, 171; infrastructure inequities, 146; inter-referencing, 145; mining sand for concrete, 175–6; pastoral capitalism, 171–2; post-suburbanization, 127–8; poverty, 171; scholarship on, 67; trends, 5
indigenous peoples, 27–8
individual choice, 24–5, 68, 190–1
Indonesia, 124
industrial and commercial suburbanization: about, 24–5, 32–3; avoidance of labour unrest, 24, 38; capital switching, 24, 197; de-industrialization, 16–18, 34; globalization, 38; historical background, 25, 32–3, 38, 66, 106; malls, 74, 193; normative ideals, 37–8; pollution away from core, 25, 28, 38; recognition in urban studies, 66; retrofitting, 74–5. *See also* Fordist and post-Fordist economy; techno-burbs
industrialized countries: historical background, 18, 24. *See also* de-industrialization; Fordist and post-Fordist economy; re-urbanization
infrastructure: about, 131–3, 148–9; capital investments, 140; critical questions, 141; cultural life, 135; decision-making, 141; density impacts, 135, 156; as a disposition, 112, 139–40, 143, 186; drivers of suburbanization, 5, 63; environmental sustainability, 132, 134, 140–1, 172; explosion/implosion dialectics, 138, 148–9; financial costs, 134, 158; formal vs. informal, 137, 139; global networks, 132; global suburbanization, 136–8, 144–7; governance and politics, 134–5, 137, 141, 142; hard vs. soft infrastructure, 136–7, 146–7; historical background, 37–8; inequities, 131, 137, 138–9, 141–2, 146; innovation, 146–7; inter-referencing, 145–6; metabolisms, 132, 137–40; metaphor of media/message, 135, 186; networks, 136–7; normative ideal, 146–7; planning failures, 142; regional importance, 131–4, 138; retrofitting, 143–9; scales of, 133, 136; scholarship on, 131; suburb as part of larger system, 136–7; technological change, 134, 137; urban planning, 134; visibility/invisibility, 142–3, 178–9. *See also* automobility; communications; mobility; public transit
infrastructure, types of: agricultural fields, 132; airports, 38, 132, 142; communications, 136; energy, 135; people as infrastructure, 137, 146; physical vs. social, 136–7; recreational spaces, 132; water systems, 132, 135, 137, 142
inner city. *See* urban
Ireland, 118–19
Istanbul, 33, **39**, 107, 115, 129, 183

Jacobs, Jane, 26
Japan,: Dejima as global suburb, 86–8, 93
Jauhiainen, Jussi, 72
Jonze, Spike, 101

Keil, Roger, 14–15
Kelly, Barbara, 72, 79
King, Anthony, 110, 122, 132–3
Kinshasa, Congo, 170–1
Kipfer, Stefan, 68
Knox, Paul, 56–7
Konvitz, Paul, 147
Korea, 165
Kovachev, Atanas, 155, 187

Labbé, Danielle, 70–1, 196
labour issues, 24, 38, 148
Lagos, Nigeria, 158
Lakewood, California: about, 84; continuity and change, 93; governance and politics, 89; homogeneity, 84, 89; model of separation, 98; myth of homogeneity and conformity, 84, 93; social class, 84, 89; Waldie's memoir, 77–8, 89–90, 93; whiteness, 84–5. *See also* California
Lamar, Kendrick, 103–4
land: about, 175; capitalism and land development, 28, 29, 197; drivers of suburbanization, 63; extended urbanization, 175; governance and politics, 184; habitable and inhabitable spaces, 170–1; industrialization, 32–3; mining sand for concrete, 175–6; 'new land,' 27–8; post-neoliberal development, 9; statistics on urbanized land, 4. *See also* boundaries; capitalism; home ownership; infrastructure; natural environment
Larco, Nico, 195
Large Technical System (LTS), 143
Larkham, Peter J., 23
Las Vegas, 109
Latin America: country-to-city migration, 37, 123; infrastructure, 139; mining towns, 177, 178; mobility, 139; scholarship on, 71; squatter settlements, 37. *See also* Brazil; Global South
Latin Americans in USA, 98, 102
Lawhon, Mary, 176
Lefebvre, Henri: about, 92–3; capital accumulation, 6; city as ideology, 43, 50; explosion/implosion dialectics, 3, 6, 10, 27, 43, 52, 69–70, 125, 138, 196; *grands ensembles* (towers), 119; Haussmannization, 33; lived experience and theory, 68; Los Angeles, 92; nature, 163; Paris as inspiration for theory, 68–9; planetary urbanization, 6; right to the city, 6, 34, 183; scholarship on, 6; suburbanization and loss, 25; suburbanization processes, 68–9, 198; suburbs as *tissu urbain*, 53; universality of urbanization, 15; urban as source of theory, 50; urban as 'virtual object,' 52–3, 189; urban

revolution, 27; *Urban Revolution*, 68
Leinberger, Christopher, 96
Levittown, 89, 110, 112, 188, 202n1 (ch. 5)
Lewinnek, Elaine, 105
Lewis, Nemoy, 85, 104–5
Lewis, Robert, 23–5, 66
Lisbon, Portugal, 101, 127, 129
Logan, John, 95–6
Logan, Steven, 203n1 (ch. 6)
London, 118–19, 165, 183
Los Angeles (Banham), 91
Los Angeles area: about, 90–2; *Anatomy of Los Angeles* (film), 91–2, 202n2 (ch. 5); centre/periphery dynamics, 90–2, 185; cultural life, 73–4; ethnoburbs, 98–9; gentrification, 7; growth dynamics, 6–7; historical background, 90, 101; inequities, 142; leader in form and function, 91; mobility, 31, 104, 142; as model for future urban forms, 90–2; policing, 101; racial and ethnic diversity, 98, 102–4; racism, 101, 142; scholarship on, 73–4; *The Sellout* (Beatty), 203n6; urban theory on growth, 66. *See also* California; Lakewood, California
Los Angeles School of urbanism: challenge to centralist tradition, 9, 64, 90, 91, 185; governance and politics, 185; influence of location of theorists, 46–7, 90
'Lovers in Dangerous Times' (Cockburn), 99

Major Collaborative Research Initiative (MCRI) Global Suburbanisms (York University), 15, 44–5, 69
malls, 29, 74, 193
Manila: *desakota* (city/rural links), 123–4, 189
Maputo, 158
Maringanti, Anant, 85
Marohn, Charles, 186–7
Marxian approach, 24
mass production. *See* Fordist and post-Fordist economy
May, Ernst, 113–14, 119
McFarlane, Colin, 141
McGee, Terry, 123–4, 155–6
MCRI (Major Collaborative Research Initiative) Global Suburbanisms (York University), 15, 44–5, 69
Mendez, Pablo, 69
Merrifield, Andy, 9–10, 177
metabolisms: about, 152–3, 167; call for new approaches, 197–8; centralist bias, 198; environmental sustainability, 152–3; extended urbanization, 177–9; greenbelts, 167, 168–9; infrastructure, 132, 137–40; natural environments, 151–3; visibility/invisibility, 93, 138, 142–3, 177
Metropocene, 173. *See also* environmental sustainability
metropolitan regions. *See* regions
Mexico City, 124
Middle East: mining sand for concrete, 175–6. *See also* Turkey

migration: arrival cities, 4, 37, 99, 177, 193; choice in, 68; country-to-city migration, 4, 37, 68, 123. See also ethnoburbs; immigrants; race and ethnicity
Miller, John, 7
mining: mining sand for concrete, 175–6; mining towns, 177, 178
Mitchell, David, 16–19, 87–8, 93
mobile homes, 188–9
mobility: about, 133–4; accessibility, 135; cyclists, 136, 145, 156; environmental sustainability, 134, 144–5, 156, 172; Euro-American model of suburbs, 23; explosion/implosion dialectics, 148–9; formal vs. informal, 139; global suburbanization, 144–6; governance and politics, 134, 141; historical background, 8, 33; inequities and justice, 139, 142, 146; interstate highways, 33; metabolisms, 138; regional importance, 131–4, 138; retrofitting, 75; scholarship on, 140, 1440; technoburbs, 147–8; technological change, 144; transit oriented development (TOD), 144–5; urban population losses, 33. See also automobility; infrastructure; public transit
model of suburbs as single family home. See single-family home suburb; single-family home suburb, as theoretical model
modernism, 113–14, 120
modernization, 112–13
Monte-Mor, Roberto, 177

Moore, Jason, 174
Moos, Markus, 69
Mumford, Lewis, 65, 114

Nagasaki Bay, Japan, 87, 93
National Center for Suburban Studies, 72
natural environment: about, 151–3; agricultural land, 166; boundaries, 137–8, 163–4; cultural life, 73; Euro-American model of suburbs, 23, 34; grey spaces, 169–70; habitable and uninhabitable spaces, 170–1; infectious diseases, 175; infrastructure, 137–8; inseparability with urban, 151–2; metabolisms, 152–3, 167–9; ostentatious displays of wealth, 56–7, 140; pollution in industrial suburbs, 25, 28, 38, 105–6; primary urbanization, 49; real and imagined landscapes, 29, 166, 168, 172; sand for construction, 175; settler societies, 110; urban elites, 28–9; urban political ecology (UPE), 151–3, 163; urbanization of, 151, 176–7. See also environmental sustainability; greenbelts
neoliberalization: about, 38–9, 199–200; call for new approaches, 199–200; global suburbanization, 123–4, 126; governance and politics, 190; greenbelts' economic value, 168; individual choice, 190–1; ostentatious displays of wealth,

38, 56–7, 140, 191; post-suburbanization, 56–7, 183–4, 190; suburban land development, 9. *See also* capitalism
Netherlands: Almere, 87–8, 93; Amsterdam, 37
newcomers. *See* immigrants; migration
Nigeria, 158
non-whites. *See* race and ethnicity
normative ideals: about, 6, 37–8, 198–9; affordability, 77; cultural richness, 77; density debates, 75, 153–4, 156–63, 175–6; dystopia vs. utopia, 9, 14, 19, 58, 101, 104, 125; failed vs. successful suburbanization, 53–4; home ownership, 77; infrastructure, 146–7; measurement of suburbs by using, 72–3; mixed use, 37–8; mobility, 75; natural environment, 77; racial integration, 77; suburban as success symbol, 8; suburban deficits, 59, 64. *See also* environmental sustainability
North America: trends, 5, 9, 15–16, 159–60. *See also* Canada; single-family home suburb; single-family home suburb, as theoretical model; United States
N.W.A. (rap group), 103–4

Oak Ridges Moraine, Toronto, 7, **162**
Ong, Aihwa, 126
Ontario: greenbelts, **162**; majority minority suburbia, 99–101;

smart growth rules, 195. *See also* Toronto area
Ortega, Amisson Andre C., 123–4

Palmdale, California, **82**
Paris: about, 119, 141; density, 119, 192; diversity of processes and forms, 119; gentrification, 183; Grand Paris, 141; Haussmannization, 8, 33, 183; high-rise developments, 119, 141; inspiration for Lefebvre's theory, 68–70; mobility, 141, 147
parks. *See* natural environment
Peck, Jamie, 57, 79–80, 190
peripheral, as term, 15. *See also* terminology
peri-urban, as term, 15, 45
Phelps, Nick, 23, 26, 45, 56, 118
Philippines: *desakota* (city/rural links), 123–4, 189
Pitter, Jay, 100
planetary urbanization, 3. *See also* global suburbanization
policing, 101, 105, 182
politics. *See* governance and politics
Poppe, Will, 192
popular culture and arts: film and TV, 74, 90–2, 202n2 (ch. 5); literature, 87–8, 93, 109–10; music, 94, 99, 101, 103–4; scholarship on, 73. *See also* cultural life in suburbs
Portugal, 101, 127, **127**, 129
post-Fordist economics. *See* Fordist and post-Fordist economy

postsocialist countries. *See* Europe, Eastern
post-suburbanization: about, 14–16, 56–7, 78–9, 189–90; boundaries, 164–5, 168–9; centre/periphery dynamics, 189–90; continuity and change, 51–2, 70, 78, 93, 124, 178, 193; diversity of processes and forms, 51–2, 56–7, 189–90; ethnoburbs, 57; governance and politics, 189–90; greenbelts, 166; intense development, 127–8; mixed land-use, 127–8; neoliberalization, 56–7; primary suburbanization, 56; research on, 78–9; terminology, 39, 43, 45–6, 56, 78, 124, 155. *See also* density; global suburbanization; retrofitting; re-urbanization
poverty: about, 94–7, 102–3; challenges in suburbs, 96; demographics in USA, 94, 102; density, 158–9; environmental sustainability, 53; extreme poverty, 94; gentrification, 183; global suburbs, 115, 158; governance and politics, 183, 191; housing estates, 191; infrastructure inequities, 141–2, 146; pejorative discourse, 96–7, 102–4; populism, 182; portrayal in *Cloud Atlas*, 16–18; racialized housing policies, 105–6; urban centres, 203n3
Prigge, Walter, 165
primary vs. extended urbanization, 49–50

private-led suburbanization, 11, 15–16. *See also* capitalism
public transit: as driver of development, 5, 31; environmental sustainability, 140–1; global suburbanization, 145; inequities and justice, 142; infrastructure, **120**; rapid transit, 104, 145; retrofitting, 75
Pulido, Laura, 105

quality of life. *See* everyday life
Quinby, Rohan, 169

race and ethnicity: about, 34, 97–8; communities of colour in suburbs, 83, 102; distanciation strategy, 34, 47–8, 65, 89, 114–15; diversity in suburbs, 83, 87–8, 95, 98; dystopia vs. utopia, 101, 104; ethnoburbs, 38, 98–9; home ownership, 32, 106–7; infrastructure inequities, 142; intersectionality, 100; majority minority suburbia, 97–101, 99–101, 104; marginal settlement spaces, 105–6; modernization and social divisions, 112–13; normative ideal, 77; pejorative discourse, 100, 104, 198; policing, 101, 105, 182; stereotypes, 182; systemic racism, 106–7. *See also* African Americans; ethnoburbs; Ferguson, Missouri; whiteness
ranch house, 111. *See also* single-family home suburb
rapid transit. *See* public transit

Redefining Suburban Studies (National Center for Suburban Studies), 72
regional transit. *See* public transit
regions: boundaries, 185; governance and politics, 39, 134, 184–5, 187, 191, 194–6; historical background, 185; importance of infrastructure, 131–4, 138; metabolisms, 152–3; 'parasitic suburbs,' 187; scale and regional governance, 39. *See also* infrastructure; metabolisms; mobility
research on suburbanization: assumptions of traditional research, 61–2; call for suburban theory, 62, 77–8, 197–8; centralist tradition, 23, 57–9; collaboration, 85; cultural life, 73–4; Cultures of the Suburbs project, 73; *desakota* (city/rural links), 123–4; ethnographic methods, 70–1, 97–8; Global Suburbanisms (MCRI) project, 15, 44–5, 69; habitation, 78; imaginaries, 78; inbetween cities, 78–9; individual voice, 77–8; influence of location of theorists, 46–7; languages, 46; literature review, 72–8; methodologies, 69–72, 77–8; multiple approaches and traditions, 46; post-suburbia, 78–9; processes vs. symptomatic value, 81; quantitative analysis, 97. *See also* theory and global suburbanization; urban studies; urban studies with suburban focus

Reston, Virginia, 77
retrofitting: about, 74–7, 192–4; change potential, 178, 193; density, 75–6; design-related activities, 74; energy renewal, 143; environmental sustainability, 75–7, 144–5, 193–4; governance and politics, 192–4; infrastructure, 143–9; inner centre vs. suburbs, 74; post-suburbanization, 56–7; privacy, 76, 194; scholarship on, 74–5; theoretical issues, 74
re-urbanization: about, 4–5, 10, 36; design-related activities, 74; European housing estates, 36, 117–18; governance and politics, 184; historical background, 18; metabolisms, 169; post-suburbanization, as term, 39; primary urbanization, 49; trends, 4–5, 10, 73–4, 117–18. *See also* density; gentrification; post-suburbanization; retrofitting
La revolution urbaine (Lefebvre), 6. *See also* Lefebvre, Henri
Ribicoff, Abraham A., 5–6
right to the city: centre/periphery dynamics, 14; historical context, 183; Lefebvre on, 6, 34–5, 59, 183; as right to the suburb, 20, 142, 183, 189, 194; suburban deficits, 59
Rio de Janeiro, 115, 146
Robinson, Jenny, 44, 67–8, 78
Ronneberger, Klaus, 43
Rothstein, Richard, 107

Roy, Ananya: on conjunctures of urban processes, 46; dimensions of urbanism, 19–20, 124–5; governance and politics, 188, 190; inter-referencing in Global South, 145; on patchwork of valorized/devalorized spaces, 49; unthinkable space, 20; worlding, 11, 20, 41–2, 49, 80, 126
rural vs. urban boundaries. *See* boundaries; greenbelts
Rutherford, Jonathan, 141

Saberi, Parastou, 68
San Fernando Valley, Los Angeles area, 73–4, 165
San Francisco Bay area, 147–8
San Gabriel Valley, Los Angeles area, 98
sand mining for concrete production, 175–6
São Paulo, 114–15, 123
Sassen, Saskia, 193
Savage, Bill, 105
scale: about, 79, 194–6; density debates, 160; extended urbanization, 8, 177–9; governance and politics, 194–6, 199–200; inbetween cities, 79; infrastructure as multi-scalar, 133, 136; multi-scalar everydayness, 12, 50; post-suburbia, 79; suburb as part of larger system, 24, 126, 136–7, 152, 156, 160; urban planning, 160. *See also* metabolisms
Scarborough, Ontario, 99–101
Schafran, Alex, 9, 96–7

Scotland, 118–19. *See also* United Kingdom
self-built settlements: about, 11; country-to-city migration, 37; governance and politics, 188; historical background, 116–17; modernization, 112. *See also* squatter settlements
The Sellout (Beatty), 203n6
Seoul, 165
Serbia, 127, **128**
service economy, 48, 58, 99
Seto, Karen, 42
settler societies: about, 110–11; Anglo-Saxon societal model, 110–11; consumerist capitalism, 110; Euro-American model of suburbs, 13–14, 110–12; historical background, 27–8, 29–30; home ownership, 30–2; ideology of freedom, 110; 'new' land, 27–8; private land ownership, 110; suburbanization as re-enactment of settlement, 110. *See also* single-family home suburb
Shanghai, 145
Sieverts, Tom, 39, 45, 78–9, 167
Silicon Valley, 57, 78, 147–8. *See also* technoburbs
Silver, Jonathan, 176
Simon, Robert E., 77
Simone, AbdouMaliq, 116, 126, 137, 170–1
Singapore, 145
single-family home suburb: about, 29–30, 76–7, 110–12, 154–5; as benchmark, 110–11; capital development, 154–5; colonial

roots, 111–12; consumerism, 111–12, 155; diversity, 94–6; historical background, 29–30, 76–7, 110–12, 154–5; home ownership, 30–2; infrastructure, 140–1; mass-produced housing, 154–5; ostentatious display of wealth, 140; poverty, 94–6; primary urbanization, 49; retrofitting, 76–7; in settler societies, 110–12. See also home ownership; Lakewood, California

single-family home suburb, as theoretical model: about, 13–14, 16, 23–4, 29–30; automobility, 153; bias of Euro-American model, 13–14, 16, 23–4, 62–3, 124; centralist tradition, 23; critics of cultural life, 31; density, 153–5; environmental sustainability, 153–5; fear of the Other, 65; privacy, 65, 153, 194; terminology, 45–6. See also terminology

slums and squatter settlements. See poverty; squatter settlements

Smith, C.B., Sr., 124

Smith, Neil, 33, 151–2

Smith, P.D., 34

social class: density debates, 159–63, 191; diversity in suburbs, 98; elite escapism, 114–15, 121–2; immigrant-based service economies, 48, 99; inequities and justice, 191; infrastructure inequities, 142; modernization and social divisions, 112–13; multi-family housing, 195; suburbanization as distanciation strategy, 34, 47–8, 89; urban elites, 28–9; whiteness, 83; working class, 84

social exclusion. See displacement; exclusion

social relations. See cultural life in suburbs; governance and politics; race and ethnicity

South Africa: historical background, 29–30; scholarship on, 67; single family structures, 108, 160; squatter settlements, 122

South Asian Canadians, 100

South East Asia, 123–4

Soviet Union (former). See Europe, Eastern

sprawl: about, 5; assumptions in critiques of, 160–1; call for new approaches, 199; density debates, 153–4, 156–63, 172, 175–6; 'drive till you qualify' home ownership, 32; drivers of, 5, 140–1; environmental sustainability, 140–1, 156, 172; European sprawl, 121; global suburbanization, 156; mobility, 5, 140–1; normative ideal, 156, 172; trends, 5; urban planning, 172. See also density

squatter settlements: country-to-city migration, 37, 123; density, 158; governance and politics, 188; infrastructure, 139; model of global suburbanization, 122; primary urbanization, 49; scholarship on, 68. See also poverty; self-built settlements

St. Louis, Missouri, 85, 104, 106–7
Stanilov, Kiril, 71, 192
state-led suburbanization, 11, 15–16, 20. *See also* governance and politics
Stockholm, Sweden, 22
'Straight Outta Compton' (N.W.A.), 103–4
strip malls, 29, 74, 193
suburban sprawl. *See* density; sprawl
suburban studies. *See* urban studies with suburban focus
suburban theory. *See* theory; theory and global suburbanization
suburbanization: about, 8–16, 75; bias of Euro-American model, 11–13, 16, 23, 62–3, 155–6; as boundary setting, 163–4; call for new approaches, 197; capitalism and production of urban space, 19–20, 173–5, 196–7; as central focus of this book, 10–16, 26–7; centralist tradition, 16, 23, 57–9, 198; centrifugal city, 10–11, 114–16, 121–2; characteristics, 15–16, 51; as concept, 12, 155–6; continuity and change, 51–2, 70, 75, 93–4, 178; diversity of processes and forms, 11, 13, 48–50, 51–2, 66, 75, 155–6, 188; dystopia vs. utopia, 9, 14, 19, 52, 58, 101, 125; extended urbanization, 8, 49–50, 177–9; increase in scale and intensity, 56; literature review, 72–7; methodology of this book, 11–12; modernism, 113–14; nation building, 110; primary urbanization, 49; processes vs. symptomatic value, 81; settlement re-enactment, 110; suburb as part of larger system, 24, 136–7; systemic and symptomatic critiques, 81; trends, 4–5, 15–16, 26–7; unknowability of, 7, 16–19, 58; visibility/invisibility, 142–3, 178–9. *See also* boundaries; centre/periphery dynamics; diversity; explosion/implosion dialectics; global suburbanization; governance and politics; metabolisms; normative ideals; terminology
Suburbia (film), 73
'The Suburbs' (Arcade Fire), 101
sustainability. *See* climate change; environmental sustainability
Swanstrom, Todd, 102–4, 107
Switzerland, 31, 35–6, **60**, 79, 160
Sykora, Ludìk, 71

Tallinn, Estonia, **150**
Tartt, Donna, 109–10
Teaford, Jon, 28, 43, 78
technoburbs: mobility infrastructure, 147–8; post-suburbanization, 57; primary urbanization, 49; recognition in urban studies, 66; research on, 78; Silicon Valley, 57, 78, 147–8
technology: communications, 137, 144; environmental sustainability, 144–5, 177; infrastructure and changes in, 134, 137, 144, 145; innovations in Global South, 145; metabolisms, 177
television series and suburban life, 74
terminology: about, 45–6; call for new approaches, 42, 45–6, 62,

157–8; chaotic uses of, 157–8; diversity of processes and forms, 11; hybrid terminologies, 45–6; languages, 46; marginalizing language, 198. *See also* theory; theory and global suburbanization; urban studies; urban studies with suburban focus

terminology, specific: Anthropocene, 173–5; density, 157–8; disposition, 112, 186; global, 125; metabolisms, 152–3; peri-urban, 45, 49, 54; post-suburbanization, 124; sub/urban, 11, 75, 193; suburbs, 23, 42, 45, 157–8; suburban, 15; suburbanisms, 11, 45; suburbanization, 45; urbanism, 125; worlding, 11, 20, 41

theory: about, 12–16, 41–2, 79–81; avoidance of dichotomies, 9–10; bias of Euro-American model, 11–12, 16, 18, 23–4, 62–3; call for new theory, 9–10, 42–3, 42–6, 50–1, 67, 77–8, 197–8; centralist bias, 14, 16, 23–4, 41, 67, 198; centre/periphery dynamics, 43, 53, 189; centrifugal city, 10–11, 121–2; city as ideology, 43; density, 42; diversity of processes and forms, 43–4, 67; failed vs. successful suburbanization, 53–4; global comparisons, 79–80; grey spaces, 169–70; literature review, 72–8; post-structuralism, 80; subjective processes, 17–19; systemic vs. symptomatic critiques, 81; theory building, 77–8; typology of suburbs, 67–8; unknowability of urban spaces, 7, 16–19, 58. *See also* explosion/implosion dialectics; Lefebvre, Henri; research on suburbanization; single-family home suburb, as theoretical model; terminology; urban studies; urban studies with suburban focus

theory and global suburbanization: about, 11, 41–2, 52–4, 121–2, 126, 129; avoidance of bias of Euro-American model, 44–5, 52, 62–3, 124, 126, 129; avoidance of centralist bias, 44–5, 53; avoidance of conceptual universality, 45; avoidance of dichotomies, 52; avoidance of regionalism or localism, 129; call for new theory, 12–13, 42–6, 50–1, 59, 67, 77–8, 80, 197–8; centrifugal city, 114–16; comparative theory, 52, 68, 79–80; critical questions, 46, 56, 62, 68; *desakota* (city/rural links), 123–4; disposition and agency, 112, 139–40, 143, 186; diversity of processes and forms, 42–3, 49–50, 128–9; dystopia vs. utopia, 52; geographical inclusiveness, 44–5; habitable and uninhabitable spaces, 170–1; hard vs. soft infrastructure, 136–7, 146–7; influence of location of theorists, 46–7; inter-referencing, 42, 115–16, 126, 145–6, 158; languages, 46; materialist approach, 126; multiple centralities and peripheralities, 16, 50, 53, 55, 177; non-urban theorists, 42; people as infrastructure, 137, theory and global suburbanization: about (cont.) 146; processes vs. symptomatic value, 81; suburb as part of larger system, 24, 126, 136–7, 152, 156, 160; urban political ecology (UPE), 151–3; worlding, 11, 20, 41–2, 80, 126. *See also* metabolisms; terminology; urban studies with suburban focus

The Thousand Autumns of Jacob de Zoet (Mitchell), 87–8, 93

Thrift, Nigel, 116

Tonkiss, Fran, 156, 158, 193

Toronto area: about, 99–101; gentrification, 183; governance and politics, 192; greenbelts, 7, **162**, 165; growth dynamics, 6–7; high-rise developments, 7, 95, 99, 117, 191, 192; historical background, 67, 99, 116–17; home ownership, 30, 99; immigrants, 99–100; inbetween cities, 142; inequities, 142; large-scale socio-spatial polarization, 95; majority minority suburbs, 99–101; poverty, 95; scholarship on, 67, 99–100; self-built suburbs, 67, 116–17; service economy, 99; social class, 99; *Unplanned Suburbs* (Harris), 116–17

Touati-Morel, Anastasia, 192

transportation. *See* airports; automobility; mobility; public transit

Trump, Donald, 181–2

Turkey: high-rise developments, 5, 37, **40**, 115; sand mining for concrete, 175–6; state housing programmes, 5. *See also* Istanbul

Tzaninis, Yannis, 87–8, 90

United Kingdom: about, 35, 118–19; Anglo-Saxon societal model, 110–11, 118–19; colonial roots of single family home, 111–12; consumerist capitalism, 110; diversity of processes and forms, 118–19; Euro-American model, 31; governance and politics, 182; greenbelts, 165–7; high-rise developments, 118–19; historical background, 29–30, 111–12, 165; home ownership, 31; housing estates, 35, 117; ideology of freedom, 110; private land ownership, 110; single-family home suburb, 13, 111–12; social class, 35

United States: Anglo-Saxon societal model, 110–11; automobility, 140–1; capitalism, 184; demographics, 94–5; dystopia vs. utopia, 19, 101, 104; environmental sustainability, 160–1; focus of urban studies, 71–2; industrialization, 32–3; political shortchanging of suburbs, 184; poverty, 94, 102–3; race and class, 19, 95; retrofitting, 76–7; as settler society, 110–11; statistics on single unit detached housing, 159–60; suburban and rural votes, 181–2; systemic racism, 106–7; Trump's election, 181. *See also* African Americans; Ferguson,

Missouri; Fordist and post-Fordist economy; Los Angeles area; single-family home suburb
unknowability of urban field, 7, 16–19, 58
Unplanned Suburbs (Harris), 116–17
UPE (urban political ecology), 151–3, 163, 167–8. *See also* environmental sustainability; natural environment
urban: about, 3–7, 15–16, 19–20, 174–5; capitalist production of urban space, 19–20, 173–5, 196–7; centralist tradition in urban studies, 23, 57–9, 63–4, 198; centrifugal city, 10–11, 114–16, 121–2; characteristics, 5–7, 15–16, 203n3; dystopia vs. utopia, 19; elite escapism, 114–15, 121–2; extended urbanization, 8, 177–9; growth dynamics, 6–7; Haussmannization, 8, 33, 183; historical background, 27; homogeneity vs. diversity, 98; inevitability of global urbanism, 20; millennials, 26, 148; normative ideals, 5, 125; polluting industries away from, 25, 28, 38; populism and protests, 182–3; suburb as part of larger system, 24, 126, 136–7; terminology, 125; trends, 4–8, 15–16; unknowability of, 7, 16–19, 58. *See also* centre/periphery dynamics; creative sectors in urban centre; gentrification; normative ideals; retrofitting; re-urbanization; terminology

urban sprawl. *See* density; sprawl
urban studies: about, 23, 61–3, 198; assumptions, 62, 198; bias of Euro-American model, 16, 57–8, 62–3; centralist tradition, 23, 57–9, 61, 63–4, 198; critical questions, 62, 68; critics of single-family home model, 31, 34; density debates, 156, 158, 160, 175–6; dichotomies, 9–10; diversity of processes and forms, 66, 188; ethnographic methods, 97–8; Global North and South, 58, 67; global regions, 67–8; historical and empirical base, 66; quantitative analysis, 97; suburban as subfield, 61, 64, 71–3, 79; urban planning, 156, 158, 160, 172. *See also* research on suburbanization; terminology; theory
urban studies with suburban focus: about, 71–2, 79–81; assumptions of suburban research, 61–2; call for focus on processes and forms, 62–3, 72, 197–8; call for new theory, 62, 74, 77–8, 197–8; diversity of processes and forms, 66, 188; ethnographic methods, 70–1; focus on USA, 71–2; global regions, 71, 172; literature review, 72–7; methodologies, 69–72; scholarship on, 63, 71; suburban as subfield, 61, 64, 71–3, 79; suburban deficits, 64; suburbs as 'new form of city,' 65–6; systemic vs. symptomatic critiques, 81; typology of suburbs, 67–8. *See also* research

urban studies with suburban focus: about (cont.)
  on suburbanization; terminology; theory
Urry, John, 140

values. *See* normative ideals
Vaughan, Laura, 61–2
vehicles. *See* automobility; mobility
Veracini, Lorenzo, 110
veranda-fronted house, 111.
  See also single-family home suburb
Vienna, 35, 114
Vorms, Charlotte, 46

Wagner, Anton, 91
Waldie, D.J., 77–8, 89–90, 93
Walker, Richard, 24–5, 64, 66, 80
Walks, Alan, 69, 70
Walsh, Bryan, 8
Ward, Peter M., 124
Weeds (TV series), 73
Whitehead, Mark, 173
whiteness: conflict with communities of colour, 84, 105; cultural life in San Fernando Valley, 73–4; governance and politics, 181–2; historical background, 32, 34; justification for racial exclusion, 106–7; in Lakewood, 84–5; myth of suburban whiteness, 83, 98; political mindset, 181–2; racial diversity in suburbs, 83, 95; white flight, 32, 34, 89; white suburbs, 32, 34, 89, 95, 106, 181–2
Wieditz, Thorben, 68
Williamson, June, 74
worlding, as term, 11, 20, 41–2, 80, 126. *See also* global suburbanization
Wu, Fulong, 45, 56, 156

Yiftachel, Oren, 170
York University, 15, 44–5, 69, **130**
Young, Douglas, 192

Zabriskie Point (film), 90–1
Zika virus, 175
Zurich, 35–6, **60**
Zwischenstadt (Sieverts), 39, 45, 78–9